CW00728659

ENCOUNTERING TECHNOLOGY

ENCOUNTERING TECHNOLOGY

THE TECH EVOLUTION I HAVE SEEN

GEORGE GERSTMAN

authorHOUSE®

AuthorHouse™
1663 Liberty Drive
Bloomington, IN 47403
www.authorhouse.com
Phone: 833-262-8899

Published by AuthorHouse 05/24/2021

ISBN: 978-1-6655-2628-9 (sc)
ISBN: 978-1-6655-2629-6 (hc)
ISBN: 978-1-6655-2638-8 (e)

Library of Congress Control Number: 2021910198

Print information available on the last page.

This book is printed on acid-free paper.

CONTENTS

INTRODUCTION

I was born before War World War II. It was 68 years before the introduction of the iPhone. On many occasions, I have been asked by my grandchildren about the way technology was when I was growing up, when I was in school, and in my professional life as an attorney. It's so interesting to see their amazement when they hear about technology of my past. So many of the things that I grew up with and used are unknown to them, and they find it amazing how different technology was before they were born.

The purpose of this book is to share with the reader my observations concerning technology of the past versus the present, as I encountered it. The chapters of this book do not form an encyclopedia of technology. Instead, they are personal accounts of what I saw and how I felt regarding technological change during my lifetime.

I have always been extremely interested in technology and it has shaped my academic and professional life. When I was in grade school and in high school, I loved trying to fix radios, television sets, lamps, cameras, electronic flash units, and anything else that piqued my technical curiosity.

When it was time to decide what I wanted to pursue in college, I decided to go for electrical engineering. To that end, I graduated from the University of Illinois in Champaign-Urbana with the degree of Bachelor of Science in Electrical Engineering. Although I planned to be an electrical engineer, I had a summer job that soured me on being an engineer and I decided to take the appropriate steps to become a patent attorney. I was very interested in law so in this way I could combine my engineering with law – a perfect combination for me. I was employed as a patent examiner at the US Patent Office in

Washington DC and attended evening classes at George Washington Law Center where I graduated with a Juris Doctor degree in 1963.

For over fifty years I practiced intellectual property law. My stint as a patent examiner from 1960 to 1963 and my practice of intellectual property law from 1963 to the present was an ideal time to watch technology evolving. I had clients in practically every technical field and each day was a new learning experience. I handled patent matters before the Patent Office, the federal district courts, the federal courts of appeal, and the International Trade Commission. My professional background is the subject of an autobiography entitled *Clear and Convincing Evidence*.

During my legal career I had the opportunity to observe the technology changing from analog to digital. The "Digital Revolution," as I saw it, began about 1970, and has been referred to in Wikipedia as the "Third Industrial Revolution." The change from analog to digital applies to numerous devices described in this book, including radios, televisions, telephones, computers, cameras, calculators, watches, video games, and more.

Some of the technology described in this book is from my lifetime as a consumer. Some of the technology was encountered as a result of my professional patent, trademark and copyright activities. The reader should understand that there are literally thousands of technical items that are not described in this book. I describe many of the technical items that I personally encountered and found most interesting and that I saw evolving over my lifetime.

CHAPTER 1

School Days

IN THE ELEMENTARY SCHOOL

I started elementary school before ballpoint pens were in existence. The first time I saw a ballpoint pen was when I was in the third grade and my teacher showed it to the class. It was called an "underwater pen" because apparently it had the ability to write underwater! Ballpoint pens were not commercialized until after World War II and I probably never used one until I was in at least the fifth grade. In elementary school we used fountain pens with wooden handles. The nib of the fountain pen was dipped in ink that was in the inkwell that was built into the wooden desk in which we sat. The color of the ink was usually blue/black, a very popular color.

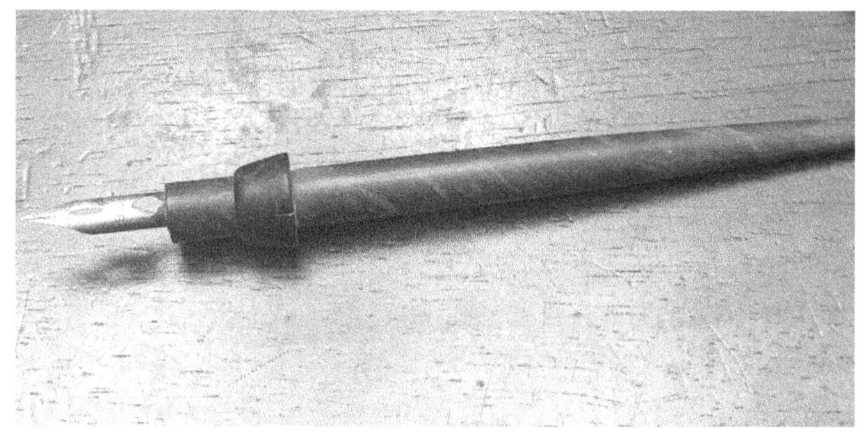

Fountain pen of the type used in elementary schools, 1940s

School desk with inkwell, 1940s

When I was in elementary school, blackboards were really black and were made of slate. The teacher and students would write on them with white chalk. The chart could be easily erased with chalk erasers but the blackboard was the main medium of written communication between the teacher and the students.

In eighth grade we had music appreciation class. We listened to 78 RPM vinyl records that were played on a hand cranked record player of the type used in the early 1900s. The teacher would crank up what we referred to as the "Victrola" in order to get the turntable revolving. A classical record would be placed on the turntable. The phonograph arm would then be placed at the outer edge of the record, and the Victrola would then play the record through the speaker that was within the console. This was an entirely mechanical system, with no electrical source used or needed.

A hand cranked Victrola talking machine for playing 78 RPM records, of the type that was in my eighth grade classroom, for music appreciation class

DANGER AT THE SHOE STORE

During the mid 1940s and early 1950s, I used to enjoy going to the Buster Brown/Austin shoe store which was near my apartment, on Austin Street in Forest Hills, Queens, New York. During the early part of those years, my mother used to walk with my brother and me to look for new shoes at the store, as our feet were increasing in size.

You might wonder, why would young boys be interested in going to a shoe store? The answer is, that the shoe store had a special machine for fitting shoes. It was a shoe-fitting fluoroscope, using x-rays to show the bones of your feet inside the shoes. What fun!

Adrian shoe-fitting fluoroscope, circa 1955

The shoe-fitting fluoroscopes were simple to use and were very popular in shoe stores throughout the United States. They were purchased by shoe stores for about $2000. Sales clerks found them to be a tremendous sales aid.

Children enjoyed climbing up on the step, and inserting their feet, with their shoes on, into the large opening in front of the machine. Their feet would be directly above the x-ray tube. The shoe salesman would flip a switch, set a timer, and the x-rays would radiate the feet. There were three viewing ports. The child could view the bones of his or her feet, with an outline of the shoes, through a center viewing port. The salesman and the mother could view the bones and the shoe outline on the two viewing ports on the other side of the machine.

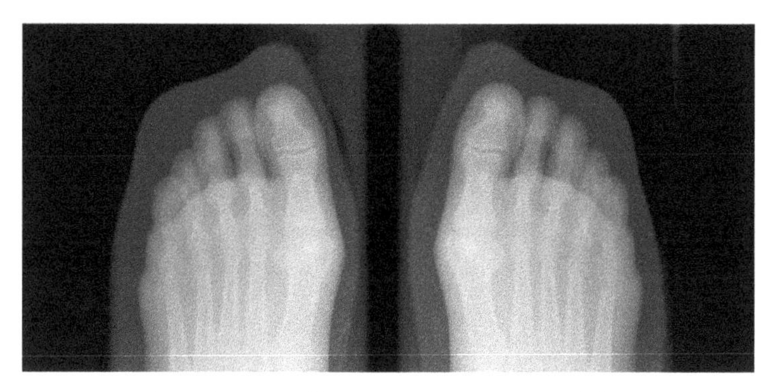

Actual x-ray of feet in shoes from an Adrian
shoe-fitting fluoroscope
(Screen shot from YouTube documentary video referred to below)

At the time, it was very exciting to see one's own foot bones in action! We would sometimes try on many different shoes so that we would have the opportunity to see our bones. Unfortunately, it was unrecognized until later that the amount of radiation emitted and received was greatly excessive and harmful. It also became questionable as to whether the fluoroscope was a valid method of fitting shoes. So the era of using x-rays as a shoe-fitting aid ended in the 1950s, with only a small portion of the present population being aware of it. An excellent YouTube four minute documentary about the shoe-fitting fluoroscope can be viewed online at www.gerstman.com/xray.htm.

Delivery Services

Before and throughout my elementary school days from 1944 to 1952 we had an electrical refrigerator in our apartment. Not every family did. At that time, all refrigerators were referred to as "ice boxes." Some of our neighbors were still using real ice boxes which were insulated cabinets containing ice for cooling the food. Blocks of ice would be delivered by an iceman to the apartment, using ice tongs to carry the ice.

An iceman's ice tongs, circa 1940

Likewise, milk was delivered in glass bottles by the milkman and placed in a metal container at the door. When the milk bottle was empty, we would leave the empty bottle by the door for the milkman to retrieve it. The bottles were sterilized, refilled and returned. There was also a man on a horse-driven cart, who would buy and sell old clothes. You could hear his cart and horse moving through the streets with him repeatedly yelling out "Buy old clothes …."

Our Tech Toys

My identical twin brother Richard and I had similar interests and we shared a Gilbert chemistry set and a Gilbert erector set. With the Gilbert chemistry set, which was a very popular item at the time, we would read the instructions and then try to perform one or more of the experiments. With the Gilbert erector set, we would build structures and usually use the electric motor to produce a movable system based on some of the systems shown in the instruction book. This was many years before LEGO blocks came into existence. We also had a Lincoln Log set which enabled us to build log cabins and the like. I'm glad to see that Lincoln Logs are still in existence because they are excellent toys for teaching young children the art of building structures from the ground.

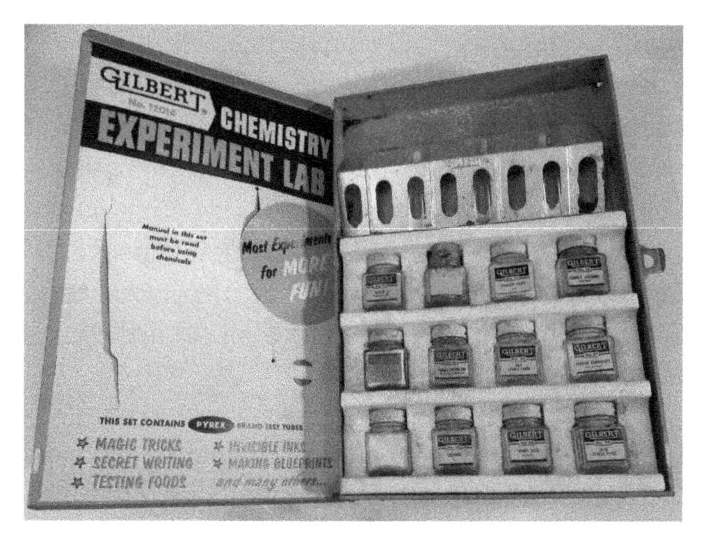

Gilbert chemistry set, 1940s

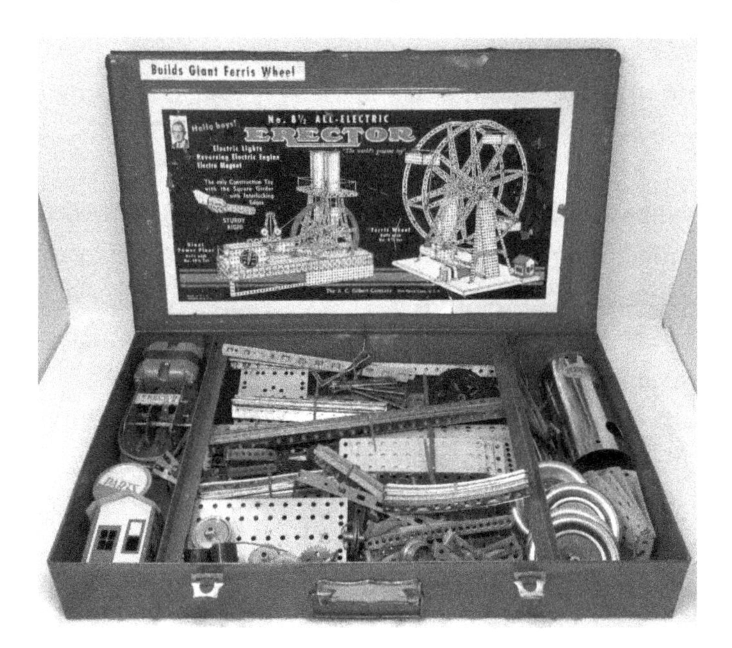

Gilbert erector set, 1940s

THE DIATHERMY MACHINE AND MICROWAVE OVEN

Throughout my elementary school years, I lived in an apartment that was two stories above a doctor's office, which was on the first floor of the building. In his office, he had an electrical device that was intended for use in relieving pain, particularly muscle pain. The device was called a diathermy machine. When in use, it was very noisy. You could hear the machine operating from outside of the office. In addition, after we purchased a television set in 1950, the diathermy machine would cause an annoying interference pattern on the TV screen when both the television set and the diathermy machine were in operation.

The diathermy machine used high-frequency electromagnetic waves to promote a "deep heating" in the body. The heat was not applied directly to the body. Instead, the microwaves generated by the diathermy machine caused the body to generate heat from within the body. It was a type of dielectric heating. If this principle of heating is starting to sound familiar, it's because it is also the principle of a microwave oven!

Today the microwave oven is a basic appliance in almost all homes in the United States, I was unaware of any microwave ovens on the market while I was in elementary school, high school, college, or law school. Except for a hugely expensive Tappan model which was rare, they were not introduced to the public until Amana began selling its Radarange oven in 1967. Once introduced, the public realized how useful and important they were in the kitchen. As with any new appliances, microwave ovens were initially very expensive but technology and competition have reduced the price of a microwave oven enormously.

Amana Radarange microwave oven, circa 1977
Photo credit: Atomic Space Junk

While I was in elementary school, we had to use the regular kitchen oven for reheating food. In the 1950s, countertop electric ovens became popular, before microwave ovens were used in the kitchen. The countertop ovens could be used for baking and broiling and were useful substitutes for the main oven. They are still in wide use, and include the toaster-oven. But the speed and ease of a microwave oven is matchless.

Our Tape Recorder

Around 1955 my brother and I purchased a used Webcor reel to reel tape recorder at a nearby appliance store. It used vacuum tube circuitry and weighed over twenty pounds. To us, it was an exciting new way of recording music and voice. It used a reel of magnetic tape which would be inserted on one side of the machine, then the tape would be extended past a recording/playback head to a fixed take-up reel. You would either play what was already on the tape or record using an attached microphone. Anything that was already recorded could be erased and overwritten.

Often we recorded songs off the radio. It was also a treat to record our voices because at that time, listening to your own voice on a recording was rare. As discussed later in this book, when I started as a lawyer in a law firm in 1963, I dictated using a Stenorette dictating machine, which was a reel to reel tape recorder.

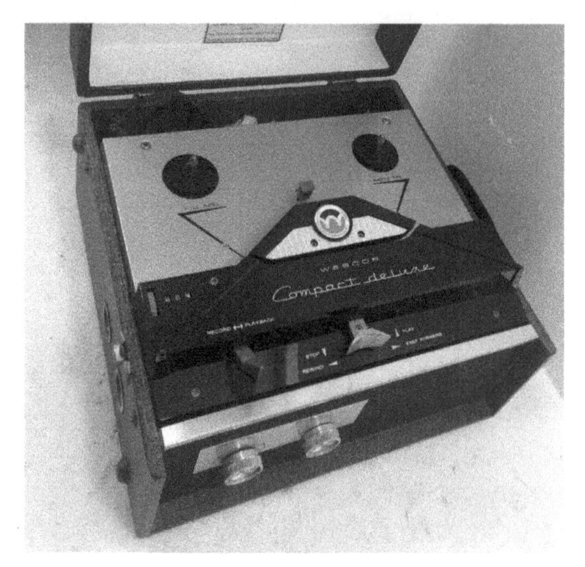

Webcor Compact Deluxe Model EP-2302 tape recorder
Photo credit: J C Clean

The Portable Record Player

During the time I was in seventh grade and eighth grade at PS 101 in Forest Hills, Queens, my brother and I took social dancing lessons at Anita Gordon's dance studio. That's where I learned how to do the fox trot, the waltz, the rhumba, and the lindy! This was 1950 and 1951. During those years and later in high school, we would occasionally attend parties where we would dance and listen to music played on a portable record player.

The record player was very simple and had a switch that allowed us to play either a 45 RPM record or a 78 RPM record. It was in the form of a small suitcase with a carrying handle. It had its own built-in vacuum-tube amplifier and speaker. These portable record players were lightweight and inexpensive, and were extremely popular at parties throughout the United States.

Vanity Fair Model 600 portable record player

CHAPTER 2

The New World Of Television

When I was a child, growing up in New York City, there was no television to watch. Television sets were not sold in any meaningful quantity to the public until about 1948 when I was nine years old. You might think: how could anyone exist without television? The answer is that it was easy: we listened to the radio!

Throughout the day and into the evening there were radio programs, many of which became television programs, such as *Superman, Gunsmoke, Perry Mason, Dragnet*, and many others. As kids growing up without TV, there were certain radio programs that we enjoyed listening to on a regular basis. Some of them that come to mind include *The Lone Ranger, The Green Hornet, Inner Sanctum*, and *The Shadow*. I listened to *The Lone Ranger* so often that I can still remember its prelude:

> "With his faithful Indian companion Tonto, the daring and resourceful masked rider of the plains led the fight for law and order in the early western United States. Nowhere in the pages of history do we find a greater champion of justice. Return with us now to those thrilling days of yesteryear, when from out of the past come the thundering hoofbeats of the Lone Ranger! Hi ho Silver, away!"

And the prelude to *The Shadow*: "Who knows what evil lurks in the hearts of men. The shadow knows……"

At the radio studio they used sound effects. For example, they would use props that would make the sound of hoof beats as well as the sound of gunshots and other appropriate sound effects. Even though we couldn't view what was happening, without knowing any better we felt that we were immersed in the action! The picture was left to our imagination and we loved it.

In 1948, very few families owned a television set because they were extremely expensive and they had very few programs. At that time, the least expensive television set was about $400 which was more than $4000 in today's money. The first television set I ever saw was in the window of a radio store on Queens Boulevard, a couple of blocks away from my apartment. I recall that it was a Philco TV with a 10 inch screen.

1948 Philco television set, Model 48-1000, with a 10 inch screen
Photo credit: The RadiolaGuy

All of the television sets used vacuum tube circuitry, a cathode ray tube for display, a large voltage transformer, and they were generally quite heavy. There were some portable television sets available, all which were also heavy. They had to be plugged in to a voltage source because they did not have an internal battery.

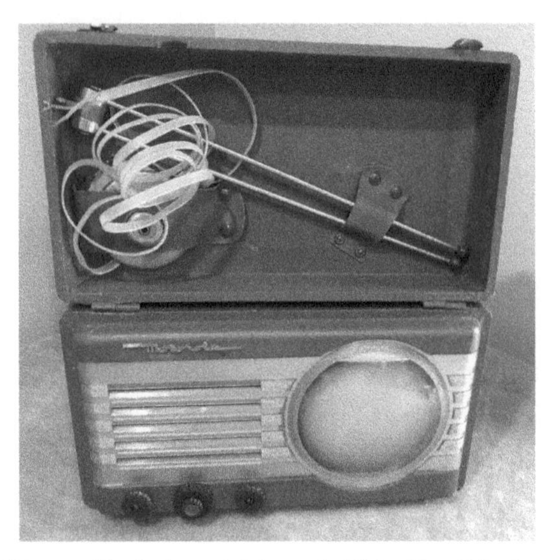

A 1948 portable (suitcase) Motorola TV with a 7 inch screen

During the 1940s and 1950s, New York City was replete with radio stores, which sold radios and also tubes and parts for radios. The radio stores had employees who could service radios. When the radio wasn't working properly, you would bring it to the radio store and they would check it while you waited. There was also a tube tester in the store for customers to use. Often the radio just needed one or more new vacuum tubes but sometimes the problem was more serious and you would have to leave the radio at the store for servicing. On the other hand, sometimes you could tell from just looking at the tubes that there was an inoperative vacuum tube. For example, the filaments of all the tubes but the inoperative one would be lit. You would make a note of the name of the tube and then buy the tube at the radio store without bringing the radio to the store.

As a selling tactic, radio stores would typically place a television set in the window facing the street so that people passing by could easily see it. However, there were only programs in the evening so the only time that the TV would be showing a program would be between about 5 PM and 9 PM in New York City. At the end of the day's programming, they would play the Star-Spangled Banner while showing a flag, and then a test pattern would be displayed until the next programming resumed. The test pattern was provided by placing a test pattern card in front of a camera and broadcasting just that. If you turned on the TV to that channel, you would see the test pattern when there was no regular programming.

RCA television test pattern

There were only four television channels in New York City including channel 2, which was WCBS (owned by the Columbia Broadcasting System), channel 4, which was WNBC (owned by RCA), channel 5 which was WABD (owned by the Dumont Television Network), and channel 7 which was WABC (owned by the American Broadcasting Company). I recall that in 1949 to the early 1950s the only time the news was on WNBC was at 7:45 PM to 8:00 PM. The news anchor on WNBC at that time was John Cameron Swayze and his show was called *The Camel News Caravan.* It was sponsored by Camel cigarettes and he would smoke Camels during the broadcast.

By 1949, some of my friends' families had purchased TV sets and they were always set up in the living room like a prize. There were more programs and after school occasionally we would watch them. The program that was probably most popular with children was *The Howdy Doody Show*, which included characters like Buffalo Bob, Clarabelle the clown, and Princess Summer Fall Winter Spring. In the afternoon, we would love to watch the cowboy movies, with Hopalong Cassidy, Roy Rogers, Gene Autry, or Tom Mix.

At that time there were no color TV's. Television worked on an analog system that is completely different from today's digital system. The display was on a cathode ray tube, the TV used vacuum tube circuitry and the picture was formed using an analog raster scan in which electrons sweep horizontally across the screen. The black and white picture had extremely low resolution compared with today's TV resolution but we didn't know any better and were thrilled to see a moving picture on the screen!

We Got A Television Set!

In 1950, the price of television sets dropped significantly. In that year, my parents purchased a 17 inch screen RCA Victor TV. We were so excited and couldn't wait for the TV to be delivered. When it arrived, it was dutifully placed in our living room for everyone to watch.

The first model TV that my family owned: a 17 inch RCA TV

By 1950 there were shows on throughout the day and in the evenings so I was often able to watch TV during part of the evening. One program that everyone in the country seemed to watch was *The Milton Berle Show*, which was on Channel 4 (WNBC) in New York City at 8 PM every Tuesday. It was preceded by 15 minutes of John Cameron Swayze and the news. During the warm weather, when the windows in our apartment building were open and we watched *The Milton Berle Show*, whenever there was a really funny line you would hear people from all over the neighborhood laughing at the same time!

One of the most popular TV shows of all time was *I Love Lucy*, a sitcom that ran on CBS throughout the 1950s. Another very popular TV show was *The Ed Sullivan Show,* which was broadcast on CBS Sunday evenings from 8PM to 9PM in New York. It was a variety show where Ed Sullivan would introduce famous stars and up-and-coming stars. What could be better for one's career than to be on *The Ed Sullivan Show*! This was the show where the Beatles were made even more famous, by being seen on this most popular show having an older audience. Certain characters appeared often, such as a ventriloquist named Señor Wences and a puppet named Topo Gigio.

At the time, the television sets had a knob for changing the channels manually. In order to avoid having to get up and walk to the television set each time you wanted to change a channel, a device was invented for allowing the channels to be changed remotely. We bought a remote control device and hooked it up to our RCA TV set.

However, this was not the type of remote control device that we use today. Instead, it was a wired system, typically with a 25-foot-long wire, in which the person watching TV could hold the switch control in his hand and with each press of the switch, a stepper motor which was attached to the channel knob on the TV, would turn the knob one channel further. A photograph of a portion of a box containing the remote control described, showing the installation, is reproduced below. From a distance up to 25 feet away, we were able to change the channels, channel by channel, without the need to approach the TV each time. However, when we crossed the living room we had be careful not to trip over the 25-foot-long wire!

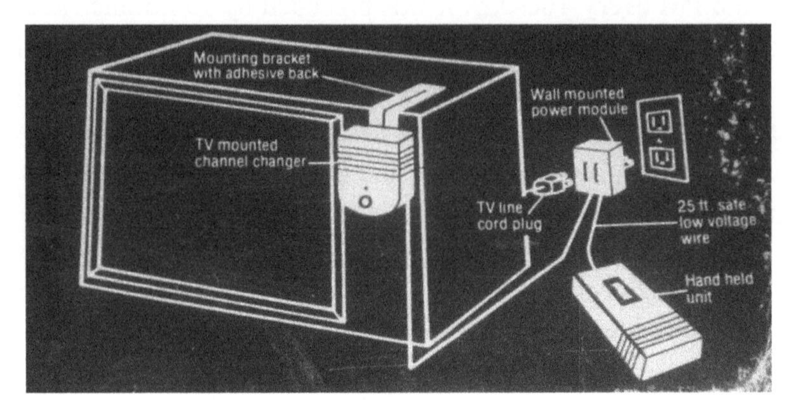

The installation instructions for the wire type remote control that were printed on the box

A wire type remote control for changing channels of a TV set, 1950s

Antennas began sprouting up all over roof tops throughout the city. You could either use an indoor antenna which was referred to as "rabbit ears," or an outdoor antenna which generally could be set up for providing better reception. Apartment buildings would typically have a roof top "master antenna" to which the tenants could connect for a fee.

The early TV sets generally had channel tuning knobs that only accessed channels 2 through 13. This was the VHF (very high-frequency) band. The UHF (ultra high frequency) band which included the channels above channel 13 was not being used in New York City until the late 1950s. It required a different antenna structure.

One type of UHF antenna that was very popular was called the "bowtie antenna." This was invented by a client of mine and I obtained U. S. Patent no. 3,363,255 for him on this invention. It was so simple yet very effective. A plastic piece simply clipped onto the VHF antenna, as shown below. My client had a small company in Illinois and sold these by the millions. Large companies from Japan were selling copies and I would charge them with patent infringement and they would stop. We kept competitors out of the market until the patent expired.

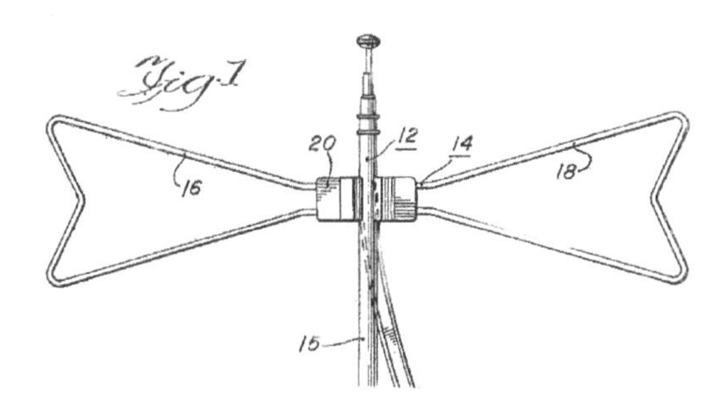

Figure 1 from the bow tie antenna patent No. 3,363,255

In the early 50s, New York City was the center of radio and television entertainment. In 1952, *The Today Show* began in New York at Rockefeller Center and it is still there. At the beginning, it was hosted by Dave Garroway. My brother and I were fascinated with television and we learned that you could see the Dave Garroway show live by simply standing on the sidewalk by the window of the studio which was right off 49th St. in Rockefeller Center. Presently Christie's auction house is in the location where the studio was while *The Today Show* has moved to a much larger studio about 100 yards east, where the public is still able to stand outside and watch the show being broadcast.

On several occasions in the early 1950s we took the subway to Rockefeller Center and watched *The Today Show* with Dave Garroway from West 49th St. There was a speaker set up on the outside of the building so during the show, we could hear the audio. However, the best part was when the show ended (as I recall about 10 AM), they provided a large studio video camera inside the studio by a small stage having a stairway. Any member of the public could walk into the studio from 49th St. to the stage, and watch himself or herself on a TV monitor. My brother and I would step onto the stage as soon as the show was over. This was really exciting because you got to see yourself on a television screen when television was in its infancy and many years prior to the commercialization of video cameras.

Since New York City was the home of television broadcasting in the 1950s, we would go to the network headquarters and obtain free tickets to watch television shows live. To obtain tickets for NBC shows we would go to the "RCA Building" at 30 Rockefeller Center and for tickets to CBS shows we would go to the CBS headquarters that was at 485 Madison Avenue, near East 52nd Street. In the morning we were able to get tickets for afternoon and evening shows and one of the shows that I remember seeing was *The Price is Right*. Even though it was an extremely popular television show, obtaining tickets to see the show was a breeze.

Color Television

During the 1950s, color TV was practically nonexistent. There was a battle between CBS and RCA (which owned NBC) as to which network's color system would be adopted by the FCC for the United States. Apparently, CBS had a system which provided a better color picture than RCA, but the CBS system programming could not be seen on a black and white TV set. On the other hand, RCA's color system was compatible with black-and-white so that any programming using an RCA system could be watched on both a color TV (in color) and a black and white TV (in black and white). In 1953 the RCA system was finally chosen by the FCC as the system that would be used throughout United States. An interesting video from 1953 of an RCA announcement of the FCC approval of the RCA compatible color TV can be viewed online at www.gerstman.com/colorTV.htm.

Although Westinghouse was the first on the market in March 1954 with a color TV, only about 30 were sold. In April 1954 RCA introduced the Model CT-100 color TV, which had a 15 inch screen. It was far more successful and is generally considered the first color TV sold to consumers.

RCA Model CT-100 color TV, 1954

Color programs were rare. NBC would make an announcement, for example, that on a particular Thursday evening at 7 PM there would be a 15-minute color program. In order to see it, you would go to a store that sold television sets and they often had a color television set that would be turned on at that time for the public to watch. I remember visiting an appliance store that was showing color TV one evening. I also recall that the picture was terrible. The colors didn't register and everything looked blurry and unnatural. I decided it would be a long time before I would even think of obtaining a color television set.

The very small amount of families that owned color TV sets would usually enjoy demonstrating the color TV to their friends while color programs were being broadcast. My belief was that the popularity of color TV was slow because they were very expensive at the time, and the color picture was far from true. Many adjustments to the color using the various knobs were constantly required. Much improvement was needed.

During the 1960s and 1970s color television improved greatly and in the 1970s most programs were broadcast in color. I purchased my first color television set in the mid 1970s. It was a 15 inch Mitsubishi color television. Television sets were still analog and used cathode ray picture tubes. The picture tube on my Mitsubishi television set was an unusual one, not as deep as ordinary picture tubes, so the television set was able to sit on a library shelf. Instead of hooking it up to an outside antenna, we used the "rabbit ears" antenna built into the television set and it operated very well.

While television sets were generally increasing in size with larger picture tubes, in 1977 a British company, Sinclair Limited, developed and manufactured a small, portable television set that could fit in a large pocket. It was called the Micro Vision TV. It had only a 2 inch display but at the time the displays were still cathode ray tubes so it still had to have substantial depth. The Sinclair TV had an internal rechargeable battery, a black and white picture, and it was quite a novelty.

The 1977 Sinclair Micro Vision TV

Throughout the rest of the Twentieth century the technology of the television set did not change drastically. The broadcasts and television sets were still analog and used standard definition. However, manufacturers were using increasingly large picture tubes (up to 43 inches measured diagonally), but these large tubes were extremely deep and heavy. TV projectors were being manufactured for projecting the TV image on a large screen. Internal projection television sets in large wooden cabinets were being manufactured, although the picture was significantly dimmer than the picture viewed directly on a cathode ray tube.

The Videocassette Recorder

The next big change in television technology was the video cassette recorder (the "VCR"), which gave the public the ability to record television programs. In addition, it started an industry of personal video recording using a video camera.

The first consumer-oriented VCR that I remember was the Sony Betamax, which was introduced in 1975. A friend of mine had the first model, LV 1901, which was a console incorporating an entire television set with a 19 inch screen as well as the tape deck and controls.

Sony Betamax Model LV-1901, with built-in 19 inch Trinitron TV
Courtesy of Rewindmuseum.com

The console cost about $2500 but then Sony introduced less expensive Betamax models such as the Model SL-8600, which were more portable. They weighed about 25 pounds, and did not incorporate a television monitor or speakers. The SL-8600 was priced at $1150 and included a digital timer and pause control.

Sony Betamax SL-8600 Beta VCR
Photo credit: K.A.S. and Stuff

Just as Betamax VCRs were getting started, JVC and RCA introduced a VCR having a different cassette format. The format was called VHS. At the time, you could only record for one hour on a Beta cassette but the VHS cassettes allowed two hours of recording. I remember buying a VHS video tape recorder as my first VCR because I liked having the idea of being able to record for two hours instead of just one hour.

One program that I recorded almost every week in the late 1970s was *Saturday Night Live*, which ran for an hour and a half. By using the VHS format I did not have to stop and change tapes. I kept some of those *Saturday Night Live* tapes for many years and converted some of them to DVDs which I still have.

Sony recognized the one-hour recording time problem and in 1977 introduced new Betamax VCRs that recorded two hours on one cassette. However, VHS recorders were then introduced that recorded four hours, six hours and then even eight hours on one cassette. Even though most say that the quality of Betamax recording was better than the quality of VHS recording, VHS video tape recorders took over and the Betamax VCR became almost obsolete. Now, all video tape recorders, which were analog, have essentially become obsolete in view of digital media including DVDs and streaming.

In the late 1970s and the 1980s, video tape recorders were the most significant new home electronic item on the planet. However, an issue arose in 1976 that could have been the end of the VCR industry. Universal City Studios brought a copyright lawsuit against Sony, alleging that the use of a VCR violates the copyright laws. The lawsuit became known as "The Betamax Case," and a demand was made to stop the manufacture and marketing of Betamax VCRs. After a trial was held, the district court found that recording broadcasts using a Betamax VCR was "fair use" and denied all relief to the plaintiffs. However, Universal City Studios appealed and the Court of Appeals reversed, holding that Sony's manufacture and sale of Betamax VCRs was a contributory infringement under the copyright laws. If this decision stood, the recording of broadcasts by the public could have ended. However, Sony petitioned the Supreme Court for review of the lower court's decision and the Supreme Court agreed to review it.

Was this the end of video cassette recorders? The future of the VCR was placed in the hands of the United States Supreme Court. Fortunately for the public and the VCR industry the Supreme Court reversed, holding:

> "Accordingly, the sale of copying equipment, like the sale of other articles of commerce, does not constitute contributory infringement if the product is widely used for legitimate, unobjectionable purposes. Indeed, it need merely be capable of substantial noninfringing uses."

The Supreme Court decision was a very close (5 to 4) decision and it almost went the other way, in favor of Universal City Studios. Papers show that at first most justices wanted to affirm the Court of Appeals but Justice Stevens, who wrote the opinion, convinced four others to vote in favor of Sony. Looking back, it's hard to believe that the home recording of a television broadcast was close to being banned.

An industry that developed in the early 1980's with the popularization of VCRs was bootleg movies on videotape. At the time the movie studios were not selling videotapes of current or past movies. They were afraid that people would stop going to the movie theaters. But there were bootleggers everywhere who were videotaping movies illegally and selling them on the black market. Many of these bootleg movies were made by people videotaping them from their seat at a movie theater. They sold for about $100 each and the video and audio quality of the recordings were horrible.

It was considered very cool to have your own movie to watch on your TV. I remember that when the Chicago gangster Alan Dorfman was murdered at a Hyatt Hotel parking lot he had in his pocket a bootleg copy of *The Verdict*. Finally, the movie studios decided to sell videotapes of their movies, but it had to be several months after the movie was released.

Displaying The Video

One of the greatest improvements in television technology was the manufacture and commercialization of flat panel TV's, with the resolution in high definition. This occurred around 2000, with the improvement of technology in the production of displays of primarily three types: plasma panel displays, liquid crystal displays (LCD), and LCD displays with backlit light emitting diodes (LED). More recently the organic light emitting diode (OLED) display has become popular.

At first, the plasma displays were the gold standard. All the best flat panel television sets used plasma displays. At the time, they were perfect for high-definition, with their deep blacks, beautiful colors, great contrast, and wide viewing angle. The plasma displays were invented at the University of Illinois, my alma mater. Panasonic was the primary manufacturer.

Panasonic had a license under the patents owned by the University of Illinois and Dr. Larry Weber, the inventor. There was a patent dispute in which Panasonic's competitor, Fujitsu, filed a re-examination proceeding against Larry Weber's main patent. Fujitsu was attempting to have Dr. Weber's patent canceled by the Patent Office. In 2002 the University of Illinois hired me to handle the re-examination proceeding. I worked with Larry Weber, and enjoyed learning the technology.

Larry had an impressive background and was very famous in the field of plasma technology. Both parties filed numerous papers in connection with the re-examination proceeding and then I went with Larry to the Patent Office to argue at a hearing before the patent examiner. I was pleased that Fujitsu lost the re-examination and the Weber patent was maintained.

Over the years, LCD displays improved significantly and became less expensive than plasma. Many of the advantages of plasma displays were overcome by new LCD technology. Due to technological advances in materials and fabrication, LCD and OLED displays have taken over the market.

Similar to early television, high definition TV broadcasts were scarce at the beginning. In addition, the flat panel TVs were extremely expensive at first but the prices decreased rapidly and they are still decreasing. At first a 32 inch flat panel TV cost about $4000. They are now available for as low as $100. Even a 75 inch flat panel TV can presently being be purchased for as low as $500. All programs are now broadcast in high definition and the FCC requires that programs in the United States be broadcast digitally.

Ultra high definition and 4K (which is even slightly higher in resolution than ultra high definition) are now the norm with most modern television sets. While the major networks do not broadcast in ultra high definition or in 4K, it can be obtained by streaming content from sources such as Netflix, YouTube, Apple, Amazon, Hulu, and others. Now you can take 4K video using your iPhone camera, upload the video to YouTube, and stream your 4K video to a 4K enabled television set with amazing results. For example, a very short (29 seconds) 4K video that I took on an iPhone X can be seen online at www.gerstman.com/4K.htm. Note the exceptional quality of the video.

Even higher resolution, called "8K" is now available on certain more expensive TV sets, although there is presently only a limited amount of content to view in that resolution.

Navigating The Viewing Selection

When TV was new, in the late 1940s and 1950s, in every metropolitan area the broadcasts were transmitted over the air. If you wanted to watch television, you attached an antenna to two bolts on the back of the television set. Connecting media to televisions these days has become much more complicated.

From then to now, television sets have had RF inputs, component video inputs, composite video inputs, VGA inputs, S-video inputs, ethernet inputs, HDMI inputs, USB inputs, and others. Presently in one of my rooms I have a cable box connected to one HDMI input, an Apple TV device connected to another HDMI input, a DVD player connected to another HDMI input, and an output of my modem connected via ethernet cable to the ethernet input. Because of all the different connected devices, many people who are not technophiles find it difficult to navigate to a desired viewing option. For example, if they are watching a DVD, but then want to switch to their favorite cable channel, they find it extremely difficult and frustrating. I recognized that problem and in 2006 I invented a device for alleviating the problem. My device was a new type of remote control having a special function. In addition to the usual buttons that a remote control device has, my device had one or more "Set" buttons and one or more "Home" buttons.

I called the first Set button "Set Home 1" and I called the second set button "Set Home 2." If the user has a desired channel that he wants the television set to revert to, while watching that channel he will press the Set Home 1 button. That setting would go into the remote control memory. Thereafter, whenever the Home 1 button is pressed, the television will revert to that desired channel, even if it was using a different input at the time. The Set Home 2 button could be activated to watch something else that is desired. For example, if the user desires to watch DVDs often, he will press the Set Home 2 button while watching a DVD. Thereafter, whenever the Home 2 button is pressed, the television will revert to the DVD input. It can be seen that my remote control device provides the user with a user controlled restore function.

I obtained three patents on my invention, the first of which was US patent No. 7,224,410. After the patents issued, the invention was sold to Wistron Corporation, a large Taiwanese manufacturer.

CHAPTER 3

The Amazing Telephone

The changes in telephony since I was a child were truly amazing. In the 1940s we had one telephone in our apartment, which was conventional throughout New York City. All home phones were black dial phones looking like the model shown below. There were no push button phones or phones with different colors.

Many households used a shared telephone line, called a "party line," which was less costly that having one's own line. The "telephone number" comprised a word, an exchange number, and four digits. For example, my family's telephone number in the 1940s was BOulevard 3-1287. The "telephone number" would be dialed by dialing the first two letters and the remaining numbers.

The popular Western Electric rotary dial telephone, 1940s

The only telephone numbers that could be dialed directly were local numbers. For example, I lived in Queens and the only numbers that could be dialed directly were numbers that were in the borough of Queens in New York City. Everything else would be a long-distance call for which an operator was required. You would have to dial the last number on the dial which was designated "Operator O" and an operator at a switchboard would answer. At the time, the operators were required to speak very clearly without any discernible accent. You would tell the operator the number of the long-distance call and she would call a rate and route operator to determine the cost of a call and how it should be routed. I say "she" because essentially 100% of the operators were women. Once the rate and route operator advised of the charge and route, the first operator would plug into a location operator who made the call to the long-distance number. The time duration of the call would be tracked because the charge for the call would be dependent upon the number of minutes of the call's duration. One of the problems with long-distance calls was that they were of low quality with much static and ordinarily it was easy to tell the difference between a long-distance call and a local call that was dialed directly.

As if regular long-distance calls were not complicated enough, many long-distance calls were made as "person-to-person" calls. You would dial the operator and then tell the operator the long-distance phone number and the name of the person whom you wish to talk to. The rate for the call would be substantially higher than a regular long-distance call but if you did not reach that particular person you would not have to pay anything, in contrast to a regular long-distance call in which there would be a charge as soon as the call was connected. With person-to-person calls, once the operator reached the long-distance number she would ask for the particular person and if that person was not available she would inform the caller and there would be no charge.

The person-to-person long-distance calling system was often gamed. One well-known method was when a person going out of town would like to provide certain information to someone at home without having to pay for a long-distance call. To that end, the person traveling would give various names to the person at home with each name signaling a message such as "I have arrived," etc. The person traveling would make a person-to-person call, ask for the coded name, and the person at home would say that that person is not available. The call would be free and the person at home would at least receive some basic information!

While the dial phones of the type shown above were used in cities and large towns, more rural areas had no dial systems and it was necessary to contact a local operator directly for any call that had to be made. In the late 1940s, my twin brother and I attended a camp in the summer located in Lake Como, Pennsylvania. At that location there were no dial phones and in order to make any phone calls you were required to directly contact an operator. The camp had a crank type telephone in which you would turn a crank and the local operator, with a switchboard, would answer. She would use her switchboard to connect to a long-distance operator in order for the call to be completed. Everyone in the town knew the local operator by name. Because she controlled the switchboard, she would be able to listen surreptitiously with her headpiece to all conversations!

A crank-type telephone, primarily used in rural areas in the 1940s

In order to obtain the telephone number of the party to be called, you would ordinarily refer to a telephone directory or, if you called from New York City or other large cities using the Bell System, you could dial 411 for directory assistance. If you dialed 411, an operator would answer: "Information." You would give the party's name to the operator and you would ask for the telephone number of the party. Upon receiving the telephone number, you could ask the operator to make the call for you or, on the other hand, you could hang up and then dial it yourself at a convenient time. You could also ask the operator to make the call for you, using long distance person-to-person calling. Until around 1980, the 411 service was free but thereafter the phone company charged for making a 411 call.

Pay phone booths were popping up all over the country. A phone booth could be found on almost every street in the most populated areas of New York City. Pay phone booths were located in rural areas, in gas stations, in general stores, and wherever there was any substantial pedestrian or vehicle traffic. In some populated areas, there would be three or four phone booths on one street, sometimes as a cluster.

In the late 1940s and early 1950s, the cost of a pay phone call was only a nickel which would be deposited into the slot on the payphone. There was a nickel slot, a dime slot, and a quarter slot. The nickel phone call was in existence for a long time but the amount changed to a dime which continued for many years. After a certain amount of time, usually three minutes, an operator would come on and require an additional payment. Long-distance calls could also be made from pay phones using the long-distance operator, by dialing 0 for operator.

Western Electric pay phone, circa 1950

In most of the United States, the oldest phone booths, which are collector items today, were made of wood, with folding doors that were formed of glass framed with wood.

Wooden folding-door phone booth, 1950s

As far as I am aware, overseas phone calls could not be made until at least the late 1950s. At that time, transatlantic cables were laid and a limited amount of calls could be made between the United States and Europe and later between the United States and other continents. In the early 1960s, I worked at a law firm in Chicago. I occasionally needed to make phone calls to Europe or Japan. The calls were very complex. First, you had to dial the operator and ask for an overseas operator. You gave the overseas operator the number that you wished to call. The call would not be made at that time but instead you would have to reserve a time at a later date. If you were lucky, you would be able to make the call the next day but often it took even longer than that. At the reserved time, you would receive a call from the overseas operator who could then make the call. Needless to say, the overseas calls were very expensive. In addition, the sound quality was very poor with an abundance of static.

In the 1950s, there was a substantial amount of talk about future push button phones but they were not available commercially until the 1960s. In the summer of 1959, I was employed prior to my senior year at college with Sylvania Semiconductor Division of Automatic Electric Corporation. Automatic Electric Corporation was a telephone manufacturing company which competed with Western Electric Company, a division of AT&T. The rotary dial on telephones was first invented by an Automatic Electric employee. We worked on semiconductor switching circuits that were intended to be used with future pushbutton phones. This was in the early days of semiconductor electronics and I was fortunate to be in a group that was experimenting with fabricating semiconductor diodes.

There were millions of rotary dial phones in use throughout the world and it would be a monumental task to change to a push button phone system. The dial phones operated by using electrical pulses. For example, if you dialed the number "3," it would produce three pulses, the number "7" would issue seven pulses, etc. The pulses were decoded to generate the telephone number being dialed. But it was decided that it would be much more efficient to use different tones for the different numbers.

What was and still is called the "touch tone" system became standard, although the earliest push button phones allowed you to choose between pulses or touch tones, depending upon your particular telephone system. The touch tone system was really a monumental invention, and an important departure from the rotary dial, because it enabled the use of smartphones and every kind of modern phone. It would be ridiculous to consider an iPhone with a dial on it!

Around the 1980s, cordless phones were introduced and became very popular. The cordless phones include a base station that is connected to a regular telephone line, and a power source for applying power. The handset includes a display and a pushbutton dialing system, a transmitter and a receiver, with a rechargeable battery for powering the handset. The handset is removable from the base station and can be carried up to a certain distance from the base station depending upon the strength of the signal and the configuration of the premises. At first the cordless phones were analog but that was found to be relatively ineffective and insecure as it was easy to eavesdrop on the transmission using a scanner. They are now digital, which improves their performance and security.

Panasonic digital cordless telephone
with base station

Once television became popular, the concept of combining the telephone with video was recognized. In 1964, AT&T displayed a new product at the World's Fair in New York that AT&T expected to be a huge success. It was called the Picturephone and it was the combination of a push button phone and a television monitor. This was really the beginning of the videophone and video calls. Picturephone booths were located in several cities, including New York and Chicago. In 1963 I joined a law firm in Chicago. The Picturephone booth in Chicago happened to be in the building where my law firm was located, the Prudential Building on Randolph Street.

Western Electric Picturephone advertisement, 1964,
Courtesy of AT&T Archives and History Center

Although the Picturephone was experimental in 1964, AT&T decided to proceed to commercialize it in 1970. However, the equipment and calls were very expensive and the caller and the receiver would each be required to have one of the Picturephones available. Also, it required that business people be motivated to look at each other as they were speaking during the business call. Unfortunately, it was a big flop. I noticed that the booth at the Prudential Building in Chicago was almost never being used.

Throughout the 1970s, 80s and 90s, many other companies produced various video phones but none was very successful. One of the biggest drawbacks was that the caller and the receiver each had to have a compatible system. This drawback was eliminated when the technology enabled video conferences via any computer. The cost of making video calls became negligible after the introduction of products such as Skype and Facetime. In addition, the resolution of the video was far superior using the Internet than prior video resolution using telephone lines.

Skype was introduced to the public in 2003 although there were relatively few subscribers at the beginning. In 2005 eBay purchased Skype. Near the end of 2005, eBay and Skype contacted me to handle some patent matters for Skype. It was my first introduction to the Skype technology, and I found it absolutely fascinating. Technically speaking, Skype is a system based on peer to peer (P2P) technology, in which each computer of a computer network can act as a server for the others. No central server is needed. I learned the technology from Skype engineers located in Estonia. Of course, we used Skype for our communication.

When we began, there was only voice communication but later in 2006 video was introduced. In order to use Skype, you would simply download Skype to any computer. You could then make a Skype-to-Skype call anywhere in the world without charge. On the other hand, you could call any landline or cell telephone in the world for a very small cost. For example, while a call via AT&T from Chicago to Australia might cost $1.50 per minute, the Skype call to a telephone in Australia would only cost about two cents per minute. The Skype-to-Skype call from Chicago to Australia would be free. Further, the resolution of the video was excellent. When the public realized how easy it was to make Skype calls it expanded quickly and there was no real need for a video phone using telephone lines.

In 2011 Microsoft purchased Skype and the technology changed slightly. Presently Skype is an app available for all smartphones. It provides a simple way to have free video calls using a Skype-to-Skype call or audio calls to any telephone in the world (land lines and cell phones) at a very low cost.

FaceTime was introduced in 2010 and Apple first showed it with Apple's iPhone 4. Anyone having an iPhone 4 or later iPhone can use FaceTime when making calls using any of Apple's IOS devices or Mac OS. While FaceTime is a wonderful, simple and reliable video and audio communication system, its main detriment is that it only can be used on Apple devices. Unlike Skype, Facetime does not connect to the regular telephone system nor can it be used with Android devices or any other platforms.

Another videoconferencing system that has become very popular more recently is Zoom. It is used extensively for social communications and it can be used with all types of computers and smartphones. Likewise, Microsoft Teams is a popular video conferencing platform. Zoom's business was greatly enhanced by the coronavirus pandemic. By mid-2020 everyone seemed to be "zooming" for both pleasure and business.

CHAPTER 4

The Telephone Becomes Mobile

MOBILE TELEPHONES BEFORE CELLULAR

For many decades the telephone was always connected directly to telephone lines (the Public Switched Telephone Network) for normal telephone communication. During the 1940s the car telephone came into existence. It was rarely seen even in the 1950s and 1960s, and it was considered extremely prestigious to have a telephone in your car.

A car telephone was very expensive to lease and it used a large and heavy transceiver usually positioned in the car trunk. Originally the equipment weighed as much as eighty pounds. The telephone handset and dialing apparatus were located by the driver's seat and an antenna was connected to the outside of the car. The antenna was typically 19 inches long. This was prior to cellular telephone systems and it was often referred to as a "mobile radio telephone." In the 1970s more compact models were developed and the entire unit could be positioned near the driver's seat.

Motorola Car Telephone Model TLD-1100, 1964

Prior to 1964, a call had to be made through an operator and only one party could be talking at a time. In 1964, the IMTS (improved mobile telephone service) was introduced and calls could be dialed directly and full duplex operation was available in which both parties could talk simultaneously. The Motorola Car Telephone Model TLD-1100 shown above was a popular IMTS phone.

Until cellular telephone service was available around 1980, it was relatively difficult to make a call over a car telephone. Calls had to be made over a small number of designated frequencies or channels. Even in large cities such as New York, there were relatively few channels. When a call was made on a particular channel, that channel did not become available for any other call until the first call was completed or unless the caller left the range of the base station, at which time the call would disconnect. It was typically difficult to get an outside line. All telephone conversations were public and could be overheard by anyone using a scanner and tuning into the channel being used. Further, the sound quality was very poor and there was a significant amount of static.

In about 1972, a client picked me up at my office in Chicago and drove me in his car to his box manufacturing factory in the Chicago suburbs. He had a car telephone and attempted to make a call during our ride but throughout the entire travel time he was unable to obtain an outside line for the call. He told me that the only times when it was easy to make phone calls were at nighttime and on weekends.

Citizens Band Radio

During the 1970s, I would see many cars and trucks with very long (up to 8½ foot) antennas extending from the vehicles. These antennas were for their citizens band radios ("CB radios"), one of the most inexpensive and simple to use mobile communication systems that became widely popular throughout the United States.

Technically, it was not a telephone system. There were no telephone lines or any telephone infrastructure involved.

CB radio is a two-way communication system. It's a broadcasting system. Any member of the public can buy and use a CB radio. No licenses are required. No telephone numbers are used. You simply purchase a CB radio, a microphone and the CB antenna and you now have two-way communications with other CB radio users. Hand-held CB radio transceivers are also available. However, CB radios do not permit the duplex operation in which both parties can talk simultaneously. Instead, you speak and wait for a response. Your audience are other travelers or truckers with CB radios, within about a three-mile to ten-mile radius.

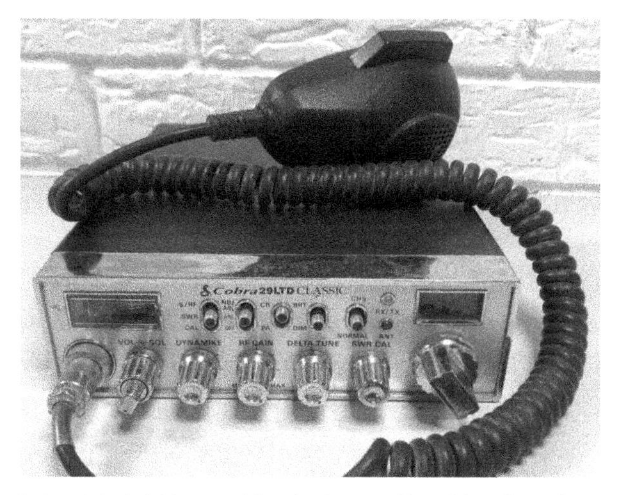

Cobra Model 29LTD Classic CB Radio with microphone

In the mid 70s you would see these CB antennas everywhere. Most of the public does not realize that CB radios have continually been used even up to the present. In particular, some long-haul truck drivers are still using them for highway information.

CB radio broadcasting works on part of the 27 MHz band which was assigned by the FCC as the citizens band. There are 40 channels within that band. Certain of the channels have primary uses. For example, channel 9 is primarily used for emergency communications. Channel 13 is the marine/RV channel. Channel 19 is a major highway information channel used by truck drivers and by travelers.

Many clubs and organizations have been formed around CB radio usage. One of my former trademark clients, React International, is an emergency responder using CB radio. It is an organization of volunteers who monitor emergency channel 9. They aid in reporting and providing communication during disasters and many other events where safety is involved. They provide rescue support and aid law enforcement.

Conversing over CB radio is unique. You have to understand that CB radio is a trucker culture. It is important that you talk like a trucker, preferably with at least a slight southern accent! There is a lot of CB radio lingo. Instead of saying "yes," you'd say "10-4." A cop is a "bear." If there are no cops around, it's a "clean shot." The gas pedal is called a "hammer." Your CB name is your "handle." If you want to know where someone's location is, you'd ask: "What's your 20?"

During the 1970s, CB radio became part of the popular culture. One of the most popular and entertaining songs concerning CB radio included a simulated CB conversation using CB lingo. This song is called "Convoy" sung by CW McCall. You can see and hear the lyrics on a YouTube video that can be accessed at www.gerstman.com/convoy.htm.

We Now Have Cellular

The cellular network system, developed in the late 1970s, was a great improvement over the earlier mobile telephone system. It has changed the course of human communication, with cellular networks located all over the planet. Cell stations can be easily recognized by the cell towers, which in high volume areas, are in close proximity to each other.

A typical cell tower

Cellular networks are ideal for mobile communications because as the user moves from a first location to a second location, there will be no interruption during this movement. Even though the frequencies of adjacent cell stations are different, the cellular phone or whatever cellular item is being used, will be switched automatically to the new frequency via what is called a "handoff." That is what makes mobile communication using a cellular network so simple – there is no need for the user to have to switch channels manually and the cellular user can simply continue to communicate smoothly even though the user is moving over a large distance.

Any communication equipment, including car phones, cell phones, tablets, computers, etc. can be connected to a cellular network. At first, it was the car phone that was the communication device ideal for connecting to a cellular network. As discussed above, in the 1970s car telephones were relatively rare, expensive and difficult to use. By 1980, AT&T began experimenting with cellular car phones in Chicago. There were 200 people selected for the experiment. A friend and client of mine was one of them. One day I drove with him to a Cubs game at Wrigley Field. By the driver's seat there was a Motorola cellular car phone with wires extending to the trunk of the car. In the trunk was a large and heavy Motorola transmitter and receiver with two antennas positioned on the trunk of the car. One of the antennas was connected to the transmitter and the other was connected to the receiver. I made a call to my home using the phone and it worked very well.

Once the cellular system was found to be viable, cellular car phones were offered to the public in the early 1980s. I purchased a car in 1984 and decided at that time to have a car phone installed. The car phone that I had installed at that time was so unusual that I cannot find a record of its existence anywhere. It was a Western Union black cell phone. The handset looked like a handset from the 1940s and it was positioned on a base cradle carrying the push buttons for making a call. The handset and base were connected near the driver's seat and there was a giant transceiver positioned in the trunk of a car. The transceiver was so large that there was essentially no room left in the trunk for anything else. A single antenna for both transmitting and receiving was attached to the trunk of the car. The car was a soft-top convertible so the antenna couldn't be attached to the rear window or the top. I remember being uneasy about having a hole drilled in the trunk of my new car for connecting the antenna to the trunk. At the time, magnetic mount antennas were available but they were less reliable than fixed mount antennas.

A vintage Motorola cellular car phone with power supply and transceiver base

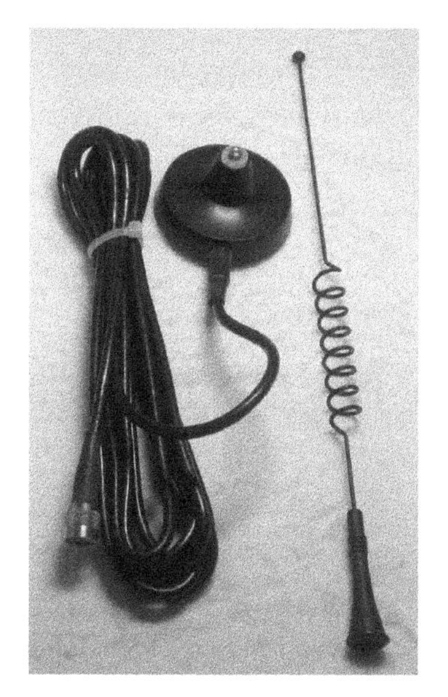

A magnetic mount cellular antenna for a car

In 1984, cell phones were just beginning to become popular in the United States, and the overwhelming use of cell phones was in cars. At least in most major cities throughout the United States, it was now easy to make telephone calls from your car using a cell phone connected to a cellular network. You could always tell which cars had cell phones because of the cell phone antenna extending from the trunk, window, or the roof of the car.

One of my clients, Mobile Mark, switched its business model from large satellite antennas to cellular antennas in order to be able to take advantage of the expansive use of cellular antennas. Needless to say, my client became a major force in the market for cellular antennas in United States. For many years the engineers at Mobile Mark invented different types of cellular antennas and I obtained numerous patents for Mobile Mark on these inventions.

Another client of mine, Security Systems, Inc., invented and developed a new alarm system use for cellular phones. This alarm system for residential and commercial use incorporated a cellular phone for communicating with an alarm monitoring station or directly with the police. Previously alarm systems, such as burglar alarm systems, were connected by tangible telephone lines to an alarm monitoring station. When an alarm condition occurred, a signal was transmitted via telephone lines to a monitoring station or a police station. In some installations, a dedicated telephone line from the premises to the police station was used.

The telephone line was an essential link for providing the alarm signal. However, a knowledgeable burglar could cut the line and simulate the voltage, which would prevent the monitoring station or police station from receiving an alarm communication. My client's invention was the use of a cellular phone which would communicate with the monitoring station or police station when the alarm condition exists. This over-the-air alarm communication obviates the problems that would occur if the telephone line were cut. The reliability of cellular networks made this invention successful and it is now used in typical alarm systems throughout the United States. Security Systems installed this alarm system in my home and I felt very safe. It is the subject of now expired United States patent no. 4,577,182.

While during the 1980s most cell phones were attached to or built into cars for the use as car telephones, there was a need for more portability. To that end, a portable cell phone, having a combined transceiver, handset, power supply, battery and antenna was developed and sold primarily by Motorola. Because of its bulk it was commonly referred to as "The Brick." It was very expensive ($3995) and it had a much lower power output (0.6 watts) than car phones (three watts). Because cellular networks were not as efficient as they are today, this relatively low wattage made it more difficult to complete calls as compared with calls with the more powerful car phones. The Brick was the first cell phone that could be easily carried by the user and it really started the massive use of mobile communication outside of the car.

Motorola DYNA T-E-C portable cell phone,
sometimes referred to as "The Brick," circa 1987
Photo credit: Niphon Chanchinda

There also was a need for a lower cost and more powerful cell phone. In the late 1980s, the bag phone was developed and commercialized. The bag phone was really a version of the car phone with the transceiver, handset, power supply and antenna combined and fit into a canvas or leather carrying case. There was an electrical cord from the power supply for connection to a car battery through the cigarette lighter. Instead of connecting to a car battery the user could purchase a 12 volt lead-acid battery for adding to the carrying case. The bag phone was particularly useful for boaters, campers, and construction sites, where a cellular phone having the same three watt power as the car phone could be easily transported.

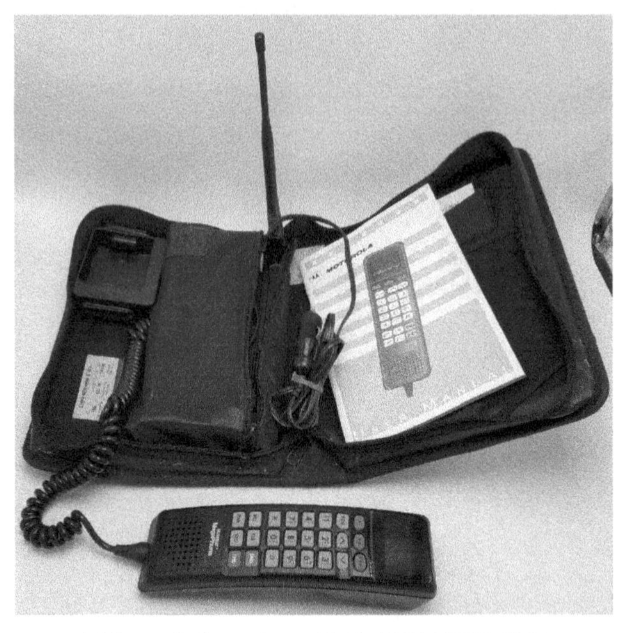

A Motorola bag phone Model 360 SCN2500A
Photo Credit: Matt Conlon

By the end of the 1980s, the size of portable cell phones was greatly reduced. In 1989, Motorola introduced the MicroTAC portable cell phone. It was a partially folding phone and the first cell phone that was small enough to actually fit in a shirt or pants pocket. In the same year, Philips introduced a very small portable cell phone which I purchased. It was slightly larger than the Motorola MicroTAC but was much smaller than the Motorola brick phone, and it easily fit in my jacket pocket. It had a 0.6 watt output. At that time, such small portable cell phones were quite rare due to their cost but I found it to be extremely useful in my law business.

Motorola MicroTAK Elite, 1994

I carried my Philips cell phone with me everywhere throughout the day. While almost everyone would have to go to a payphone to make a call when outside of the office or home, it was so easy to merely take the phone out of my pocket or briefcase and make or receive a call. I did a lot of traveling at the time, and I found it was extraordinarily useful. I even took it with me on my Sea-Doo personal watercraft so that I would have telephone communication while I was jet skiing around Lake Michigan. I kept the Philips phone in a plastic case in the personal watercraft and I always felt safer having the phone available.

I believe that it wasn't until around 1995 that the small portable cell phones began to get very popular. In 1996, Motorola introduced the StarTAC portable cell phone which at the time I thought was the most practical and beautiful new product on the market. I purchased one immediately and loved it.

Motorola StarTAC cell phone with charging cable attached, 1996

It was extremely small, enabling it to fit very easily into the pocket of my jeans. It was also very light (about three ounces) and easy to use. It was the first flip phone and it was in the form of a clamshell. When folded the display and all of the keys were fully protected. It had a telescopic antenna at a top corner that normally did not need to be extended during calls. It enabled text messaging and it could be set to vibrate instead of ringing. The StarTAC cell phone was a huge commercial success for Motorola.

Apple Introduces The Iphone

In June 2007, a device that has been referred to as the greatest invention of all time began to be sold. It was the iPhone. It revolutionized the world. It was a pocket-size computer. It was connected to the Internet. It didn't look like any other phone: it was rectangular and flat. It didn't fold. It was sleek and simple. It didn't have a physical keyboard including distinct buttons that had to be pressed. Instead, the entire front screen was formed of glass and was a touch screen with a virtual keyboard. Although that seems commonplace on smartphones now, in 2007 it was novel.

The iPhone that was first sold in June 2007 is commonly referred to as the iPhone 1. It was 4 ½ inches high, about 2 ½ inches wide, and slightly less than one half an inch thick. At that time, the amount of functions found in a single pocket-sized device was amazing. It had some basic apps including phone, calendar, photos, stocks, weather, clock, calculator, notes and settings. This made it a telephone, telephone directory, calendar, still camera, photo viewer, clock, and calculator, all in one pocket-sized device. In addition, it had voicemail including visual voicemail, email, a Safari web browser, text messaging, a music player, a YouTube app and a Google Maps app.

The iPhone 1, 2007

At first, in the United States AT&T was the only cellular provider for the iPhone. iPhones could be purchased only at an Apple store or at an AT&T store. People camped out for days in advance in order to be early purchasers of an iPhone. On the first day of sale, the lines were around the block or longer at each store. I decided to wait about a week in order for the lines to calm down and then I purchased an iPhone 1 at an AT&T store. I was handed a fresh black box containing an iPhone together with a new SIM card carrying my cellular information. I brought it home and I was amazed how easy it was to set up the iPhone.

There was no app store app when the iPhone was first introduced. In order to obtain additional functions using third-party apps, you had to remove Apple's software restrictions by what was referred to as "jailbreaking." But that would also remove Apple's warranty and could possibly cause other problems. In 2008 Apple introduced the app store app in an iPhone update, and third-party apps approved by Apple became available. Additional apps make the iPhone more versatile and there are presently over 2 million apps.

There was no video camera on the iPhone when it was first introduced. A video camera was not incorporated into the iPhone until 2009, and it has now become one of its most advanced and exciting features.

While there were hundreds of new functions available on the iPhone 1 in 2007, one function that I considered revolutionary that is now commonplace was as follows. If you needed a special service such as a private medical plane in an emergency, you could simply enter the search term "private medical plane" in the Maps app or in the Safari browser and a list of selected air ambulances would become available. You could then click on the phone number link and speak with the appropriate person. You could make the necessary arrangements over the iPhone and continue with follow-ups over the iPhone until or after the plane arrived. In this manner, the entire transaction from finding an appropriate air ambulance to the completion of the trip to the hospital could be made using the iPhone. Being able to use a portable pocket-size device as a combination telephone/computer was a real breakthrough at that time.

Although I have had many new iPhones since my original iPhone 1 in 2007, I have kept my iPhone 1 as a memento. I am amazed that at the time of this writing, the battery is still strong when charged and the iPhone1 works perfectly!

The iPhone 1 as sold in 2007, with charging cable, earphones and charging stand

The huge success of the iPhone attracted other mobile phone manufacturers around the world, who made smartphones that are generally similar in appearance to the iPhone. Apple's main competitor is Samsung, which has a greater share of the smartphone market than Apple. Other smartphone manufacturers include Huawei, LG and Motorola.

Although the flip phone had become almost obsolete, in 2020 Samsung introduced two new folding phones in which the entire front surface is a touch surface carrying an image. In a typing mode, part of the image is a virtual keyboard. In a photo or video mode, the phone is unfolded to reveal a double size image surface.

Samsung Galaxy Fold
Photo by Mike Baumeister on Unsplash

CHAPTER 5

How They Shrunk The Computer

My first encounter with a digital computer was in 1959, when I took a computer course as an electrical engineering student at the University of Illinois. Digital computers were very rare at that time but the University of Illinois had a top computer program and I was lucky to be able to enroll in it. There were two digital computers at University of Illinois: The Illiac I and the IBM 650. The Illiac was built at the University of Illinois and was older than the IBM 650.

The University of Illinois was famous for having the Illiac computer. They kept it in operation 24 hours a day. It was constantly being used, with lights always flashing. The computer was massive: it weighed ten thousand pounds and had 2,800 vacuum tubes.

The Illiac I computer at the University of Illinois.
Photo courtesy of the Board of Trustees of the
University of Illinois

The input to the computer was a Teletype machine with Teletype tape. The output from the computer was also a Teletype machine in the form of a Teletype teleprinter. There was nothing that today would be considered an "operating system," so we had to program every operation. All the instructions were entered digitally using the Teletype tape. The tape was wide enough to handle five aligned punched holes. The first hole corresponded to a one, the second hole corresponded to a two, the third hole corresponded to a four, and the fifth hole corresponded to an eight. As an example, to enter a three, the first hole and the second hole would be punched by the Teletype machine.

The amount of storage was infinitesimal compared to the amount of storage used today. There was no transistor storage at the time. The storage and memory consisted of thousands of ferrite rings which were ring magnets that were threaded together. The magnets were either in a magnetized or non-magnetized state, depending on whether they were storing a 1 or a 0.

Ferrite core with 14 X 20 magnetic ferrite rings, 256 bits memory

After learning the basic tasks in programming the Illiac 1 computer, I was taught how to program the IBM 650 using its operating program named SOAP (which was related to FORTRAN).

This operating program enabled me to use fewer instructions in order to complete a task. The IBM 650 was a more modern digital computer. It was the first mass produced digital computer. The 650 was IBM's first commercial business computer. It also was a large general purpose computer with flashing lights and vacuum tubes.

IBM 650 computer at Texas A & M University
Photo credit: Cushing Memorial Library and Archives, Texas A&M

Instead of using Teletype tape, the IBM 650 used punch cards which became typical with IBM mainframe computers. A keypunch machine having a typewriter input was used to provide the punched holes in the cards and each instruction required a separate card. I would compile a large stack of cards which would then be fed into the computer input.

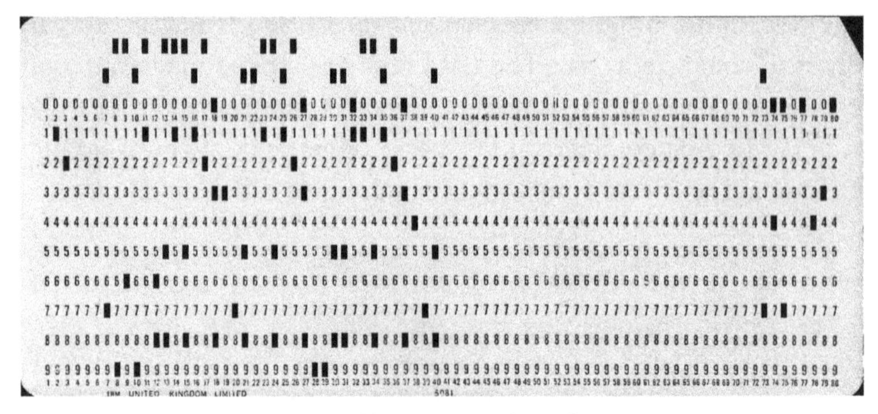

An IBM-type punch card
Photo credit: Pete Birkinshaw from Manchester, UK / CC BY 2.0 (https://creativecommons.org/licenses/by/2.0)

At the time the punch cards seemed to be everywhere and an expression that became famous with punch cards was “Do not fold, spindle or mutilate.”

After graduating from the University of Illinois with an electrical engineering degree, I was employed as a patent examiner by the US Patent Office which at that time was at the Department of Commerce building in Washington DC. My employment extended from July 1960 to August 1963, during which time I also attended evening law school at George Washington Law Center.

One of my duties as a patent examiner was to classify patents relating to semiconductor devices. To that end, I had access to the computer at the Department of Commerce, which was an IBM 305 RAMAC digital computer. I, personally, did not program the RAMAC computer but I worked with the computer staff.

IBM RAMAC computer with two disk drives
in the foreground and console in the background

The computer used vacuum tubes, weighed over a ton and was very slow. It was the first computer with a magnetic random access disk drive, like the hard disk drives that are presently being used. There was a stack of 50 discs that you could view through a window and actually watch the head accessing the disc. The stack of disks had the appearance of a stack of records on an old jukebox.

The RAMAC disk drive, with a stack of 50 disks.
Photo credit: vnunet.com / CC BY-SA 2.5
(https://creativecommons.org/licenses/by-sa/2.5)

Although the RAMAC computer has been obsolete for many years, it really was the first device to use IBM's invention of a hard disk drive. Hard disc drives, such as those manufactured by Seagate and Western Digital, are very popular now for data backup and other data storage.

Inner view of a contemporary Seagate 3.5 in.
hard disk drive, stack of three disks
Photo credit: Eric Gaba, Wikimedia Commons user Sting / CC BY-SA 3.0 (https://creativecommons.org/licenses/by-sa/3.0)

The evolution of the shrinking computer continued in the mid 1960s when the minicomputer was introduced to the market. The minicomputers were smaller and substantially less expensive than the IBM mainframe computers. The leading manufacturer of minicomputers was Digital Equipment Corporation, often referred to as "DEC," which was formed by ex-employees of IBM. Another large manufacturer of mini computers was Data General Corporation, which was formed by ex-employees of Digital Equipment Corporation.

In 1973, I became very familiar with Data General Corporation. Data General was sued in federal court in Chicago by Electronic Processors, Inc. for alleged patent infringement and I was hired as lead counsel to defend Data General. The patent in suit, US Patent 2,887,674, related to a cassette tape system for storing binary data using a pulse width technique. My primary defense was that the plaintiff's patent was invalid because the invention was shown in a prior art publication. The case was tried before federal Judge Joel

Flaum who agreed with my defense and invalidated the plaintiff's patent.

A transparent cassette tape

At the time of the trial, cassette tape storage was not very popular. Floppy disks had arrived in the market and were being used for portable storage. However, when Apple introduced the Apple computer, both floppy disks and cassette tapes were used and the cassette tapes became popular again.

Four Imation floppy discs

The Transistor

In my opinion, the shrinking size of the computer was primarily due to the invention of the transistor. The transistor was one of the most important inventions in history. It was invented in 1947 at Bell Labs although transistors did not become popular until the mid 1950s. One of the inventors of the transistor, Dr. John Bardeen, was a professor at the University of Illinois while I was there. He was awarded the Nobel prize for the invention of the transistor and obviously was highly respected by the electrical engineering faculty.

Transistors are semiconductor devices, able to amplify electrical signals and switch electrical signals. These features enable them to become an extremely effective substitute for vacuum tubes. An important feature of the transistor is that when placed in certain arrays, transistors can become logic gates, enabling binary 1's and 0's to be generated, stored and processed. This is ideal for digital computers which utilize binary 1's and 0's in their operation. At first, transistors were large with each transistor being the size of a coin. However, while I was in college in the late 1950s there was extensive research on the miniaturization and the microminiaturization of transistors.

Assorted discrete transistors
Photo credit: Transisto at English Wikipedia / CC BY-SA 3.0 (http://creativecommons.org/licenses/by-sa/3.0/)

The invention of the MOSFET (or MOS transistor) significantly changed the landscape and enabled millions of MOSFETs to be fabricated on a single silicon chip. This invention changed the world of electronics and led to the development of microprocessors, and the miniaturization of all electronic devices that are "computer-controlled." The MOSFETs are etched on a silicon chip using photo-lithographic techniques. It's hard to believe that the iPhone 11, for example, uses a chip (Apple's A13 chip) that contains 8.5 billion MOSFETs.

I was fortunate to have experiences in the field of semiconductor technology. My first experience was in the summer of 1959, when I had summer employment as an electronic technician at Sylvania Semiconductor Laboratories in Northlake, Illinois. We were researching the use of semiconductor diodes for telephone switching circuits. The diodes were like transistors but having two electrodes instead of three electrodes, so they did not perform all of the functions of the transistor but were mainly used for switching functions. I actually fabricated the experimental semiconductor diode switching circuits by hand, starting with raw silicon.

I was introduced to the technology of fabricating integrated circuits containing MOSFETs in 2004. I was the patent expert witness on behalf of ASML, a large Dutch manufacturer of photolithographic equipment for fabricating integrated circuits. ASML was on trial for alleged patent infringement of a patent relating to a certain photo reduction technique. It was fascinating to learn how an optical lens system is able to reduce a large circuit pattern to a microminiature circuit pattern that is projected onto a silicon wafer using the optical lens system.

In 2005, I was Samsung's patent expert witness in a patent infringement case concerning Samsung's invention for a type of DRAM. A DRAM (pronounced "dee-ram") is a dynamic random access memory device primarily utilizing MOSFETs combined with capacitors to form memory cells. DRAMs are able to store binary data based on whether the capacitor is charged or discharged. For example, a charged capacitor may represent a 1 while a discharged capacitor may represent a 0.

DRAMs are often used as the main memory of computers. It should be understood that in general, the greater the amount of RAM the greater the speed of the computer. It is the RAM in the computer that is making the computer's calculations. The type of RAM used in computers today is primarily the DRAM.

Depending upon the amount of memory in the computer, a DRAM may contain billions of MOSFET transistors. The differences between the cost, speed and efficiency of the random access memory now provided by DRAMs and the old random access memory provided by magnetic disks (such as in the IBM RAMAC discussed above) are staggering!

The Microprocessor

In the late 1960s and into the 1970s, smaller computers contained stacks of circuit boards having multiple integrated circuits on each board. Around 1970 there was a major change in circuit technology. It was the microprocessor. The first commercial microprocessor was introduced by Intel Corp. in 1971. It was the Intel 4004 microprocessor and was designed to be used in an early portable calculator. In 1971 it was called a "microcomputer on a chip."

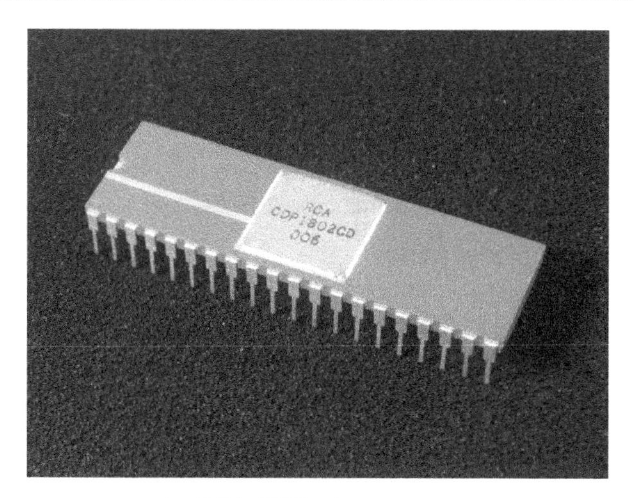

An RCA 1802 microprocessor

The microprocessor is the brains of a computer; it enables the computer to perform its tasks. Instead of requiring stacks of circuit boards having multiple integrated circuits on each board, the entire processing system for the computer can be fabricated on a single chip in the form of a microprocessor. It is the CPU, or the central processing unit of the computer. Digital data is fed to the microprocessor where it is processed according to its programmed instructions that are read from the memory (the RAM), and then the results of the processing are output in the form of digital data.

It is helpful to understand at least the basic components of the modern computer. The microprocessor, which as stated above is the brains of the computer, runs the computer's computations. The microprocessor forms the central processing unit of the computer, typically referred to as the CPU. It receives instructions from the RAM which sends the instructions as binary data. The RAM is a memory unit in which the data stored in the RAM disappears when the power to the computer is turned off. The RAM stores the data that is being used and feeds it to the CPU for processing.

You save your miscellaneous files and data onto a storage drive. The storage drive may be a hard disk drive or a solid-state drive.

These files and data are in permanent storage until you want to delete them. For example, you may purchase a particular program such as Microsoft Word and store it on the solid-state drive. When you open up the program, you see a window on your computer and the data is fed to the RAM. In order to process the data, it is fed by the RAM to the CPU.

The computer also must have a power supply for providing power to the components of the computer. The power supply, microprocessor, RAM, and other components are typically carried by a motherboard (which is a circuit board). The motherboard also can contain input and output terminals for connections to and from peripherals, such as a keyboard, a mouse, a monitor, a DVD drive, etc. The motherboard may also have extra slots for receiving other smaller circuit boards.

In a computerized electronic device, such as an arcade video game, a pinball machine, etc. the internal permanent storage device for the digital data (the "digital data" is the computer-readable code forming the program) is a ROM ("read only memory"). Often multiple ROMs are used for a very large amount of code. For example, the detailed operational program for an arcade video game would be stored in a number of ROMs, located on a circuit board within the video game. The ROMs are usually programmable and are referred to as "PROMs."

The Personal Computer

In 1974, Intel introduced the Intel 8080 microprocessor. It became the central processing unit of what is considered to be the first successful personal computer, the Altair 8800 manufactured by MITS. The Altair 8800 was released in 1975 and was primarily sold through advertisements in hobbyist magazines. It was considered a "microcomputer," and it enabled hobbyists and professionals to have and use their own personal computer, using the BASIC programming language.

Altair 8800 computer
Photo credit: Boston based collector Tim Colegrove

Prior to 1975, digital computers were considered by the public to be large and expensive devices, used in business and industry. Very few people would even consider having their own personal computer. When the Altair 8800 entered the market, it seemed unusually inexpensive for the price of a real computer. This was not a simple calculator: it was a digital computer. Yet it could be purchased in assembled form for under $1000. It cost even less when sold as a kit. I understand that the Altair 8800 computer was discussed at a hobby show which Steve Wozniak attended and it inspired him to develop the Apple computer with Steve Jobs.

Altair 8800 computer interior
Photo credit: Boston based collector Tim Colegrove

The first personal computer that I owned was the Apple II computer.

The Apple II computer with two floppy disk drives
and an Apple monitor

Because I had done some computer programming in college, I thought it would be fun to experiment with programming using the BASIC language. I also wanted to be able to do word processing. In the late 1970s, the Apple II computer was starting to get very popular, primarily in view of its simplicity in design and the concept that you could now own your own computer in the form of a device looking like a typewriter.

As shown in the above photograph, the Apple II computer had a built-in keyboard. The back section could be opened for simple manipulation of the various circuit cards that were available. You could purchase an Apple monitor as shown in the photograph, or you could attach any monitor. Peripherals such as floppy disk drives and printers could easily be plugged in.

When it was first sold, the Apple II computer used a tape cassette for data storage. Later, a floppy disk drive was designed for connection to the computer, as shown in the photograph above. Many of us remember those old large floppy disks that were 5 ¼ inches in diameter.

Now when you purchase a computer it is typical to have 16 GB (16 gigabytes) of RAM (dynamic memory). That is over 16 billion bytes of memory, contained on a DRAM. In contrast, the basic Apple II computer had only 4K (4 kilobytes) of RAM, which is only four thousand bytes of memory. If you paid extra and had 16K of RAM, that was a big deal!

A December 1977 advertisement from Byte magazine for the Apple II computer, detailing many of its characteristics, can be viewed online at www.gerstman.com/Apple2.jpg.

Because of its simplicity and popularity, many consider the Apple II computer to be the first successful personal computer. I understand that over 6 million Apple II computers were sold.

I enjoyed programming in BASIC language on my Apple II computer. BASIC was built into the computer. There were plenty of published sources from which to learn the BASIC language. Also, it seemed much simpler than the programming language that I used in college when programming the Illiac and the IBM 650 computers. With my Apple computer, I would make various mathematical programs in which numbers could be inserted to enable the computer to compute the result. I recall a program that I wrote for handling financial portfolios. I also enjoyed experimenting with word processing including editing and being able to right-justify my typing product. I remember using a word processing program called Apple Writer.

The Apple II computer was not the only personal computer in the late 1970s. The year 1977 was an extremely important year with respect to the advent of personal computers. While I recall the Apple II computer being the first successful personal computer, in the same year there were two other personal computers that were mass-marketed to the public and became very successful. One was the Commodore PET computer and the other was the RadioShack TRS-80 computer. Like Apple II, both of these computers had built-in keyboards. The Commodore computer looked very much like an ordinary electric typewriter. It didn't have a built-in monitor like the TRS-80.

Commodore PET computer with built-in keyboard

Radio Shack TRS-80 computer, with built-in keyboard, built-in monitor and two floppy disk drives

For those such as hobbyists who wanted to spend less and have a personal computer, at approximately the same time The Heath Company sold a Heathkit which had a computer in kit form. You bought a Heathkit and if you were handy, you could assemble the parts together yourself and have a working personal computer. At the time, the Heath company was well-known for selling all types of Heathkit electronic products in kit form, including television sets, radio equipment, and electronic test equipment.

The biggest computer manufacturer in the world, IBM, had nothing resembling a personal computer at the time. But after seeing how there appeared to be a large market for a microcomputer such as the Apple II or the Commodore PET or the RadioShack TRS-80, IBM decided to jump in. IBM's first microcomputer was introduced to the market in 1981 and became a huge seller. It was the IBM PC 5150 and sold for about $1600 with 16K RAM, which many considered a bargain for an IBM computer. After IBM had sold computers for hundreds of thousands of dollars and even millions of dollars, the public trusted IBM to have a top computer for personal use.

It's amazing to consider how back then, the standard was a 16 K memory, while presently the standard is a 16 GB memory which is about one million times more memory!

IBM PC 5150 computer, with built-in floppy disk drives and wire-connected IBM monitor and keyboard

While Apple kept its operating system proprietary, IBM decided to have an open-source operating system. In this manner, peripherals manufactured by third parties could easily connect to and work with the IBM computer, including printers, input drives, and anything else that could possibly be connected to a computer. In addition, software companies could develop software that was compatible and useful on the IBM computers.

This open source operating system also enabled other companies to copy IBM's operating system and to manufacture what were referred to as "clones" of the IBM computer. These clones looked like and worked like the IBM computer but were made by other manufacturers throughout the world. There were dozens of these clones on the market, most of which sold for substantially less than the price of the IBM computers.

The personal computer market really fell into the two systems: The Apple Mac OS system and the IBM system. The IBM personal computer and the clones of the IBM System were referred to as PCs and they still are referred to as such. In the 1990s in particular, PCs were far more popular than Apple computers.

In my law firm, the secretaries used dedicated word processing devices for word processing during most of the 1980s. These were computers dedicated to word processing, having a CRT monitor. Many of the steps required for formatting a page were difficult. At the end of the 1980s we began purchasing IBM clones for each secretary and every few years, the computers would have to be upgraded with new IBM clones. In the 1990s most of the computers we purchased were Dell computers.

In the early 1980s, WordStar was a popular word processing software. However, it was replaced with WordPerfect, which became dominant particularly throughout the business community. In particular, practically every law firm in the country used WordPerfect for its word processing software.

At first the WordPerfect software was on a floppy disk which was purchased for a fixed price. The disk did not carry any copy protection, so it was common for a company to buy one disk and use it on numerous computers. Finally copy protection software became prevalent and each disk required the use of a code which enabled the software to be used with only a single computer or up to a specific number of designated computers.

Although Microsoft introduced Microsoft Word in the mid 1980s, it did not become popular until the 1990s when it was bundled with Windows. By the late 1990s, Microsoft Word really had taken off and started to dominate the market in word processing software. However, most law offices stayed behind and kept using WordPerfect until about the year 2000. Presently WordPerfect has only a minute share of the word processing market.

Now the most popular word processing programs are Microsoft Word and Google Docs. Microsoft Word is primarily a program that you have to purchase with a Microsoft Office suite of programs. You can then install it on a limited number of computers. It's a very powerful program with an extraordinary number of features. I'm using Word to write this book. On the other hand, Google Docs is a free online program that you access on the Internet at docs.google.com. It's particularly useful if you want to collaborate with others over the Internet in writing and editing a document.

Until about 2000, Microsoft was primarily a software company. Its most famous products were the Windows operating systems, the Microsoft Office suite including Microsoft Word, and the Internet Explorer web browser. Certain of Microsoft's software contained code for optimizing access to the program during startup thereby reducing the launch time of the computer program.

A patent infringement lawsuit was brought against Microsoft by a Texas company named Computer Acceleration Corporation (CAC). CAC claimed that Microsoft's software was infringing CAC's US patent. CAC was seeking hundreds of millions of dollars as damages. I was hired as Microsoft's patent expert witness for the trial. Microsoft claimed that the patent was not infringed and was invalid. The trial was held before a jury in Beaumont, Texas. I did my best during my testimony to inform the jury about the technology and the proceedings before the United States Patent Office. The technology was very complex and none of the jurors had a technical background. Four of the eight jurors had never used a computer. Fortunately for Microsoft, the jury verdict was completely favorable in all respects.

Although most of the attorneys and the secretaries in my law office were using desktop computers, in the early 1990s I decided to try a laptop computer. My first laptop computer was a Toshiba laptop. At the time, laptops typically were not as powerful as desktops in that they had weaker processors, less RAM memory and less storage. Also, the laptop screen was not nearly as bright as the CRT screen used with desktop computers. However, I liked the idea of being able to carry it in a case and use it at hearings, trials, and at client conferences. I also purchased a small thermal printer which I was able to squeeze into the computer case.

As time progressed, laptops became much more desirable because the price decreased, the screen resolution and brightness increased and most of all, the weight and size decreased. It was and still is common for a laptop manufacturer to emphasize how thin and lightweight the laptop is. Further, the technology has enabled the laptops to simulate desktops by having a fast processor and a significant amount of RAM memory and storage. For example, you can presently purchase Apple's MacBook Pro with up to 64 GB memory (that's 64 billion bytes) and up to 8 TB storage (that's 8 trillion bytes). The storage medium in the better laptops is a solid state drive as contrasted to a hard disk drive.

The next type of computer that shrunk in size is the tablet computer, usually simply referred to as a "tablet." Although there were certain tablet computers in the 1990s, the tablet that became hugely popular was Apple's iPad, which was introduced shortly after the iPhone. In a sense, the iPad was really a very large version of the iPhone. It was primarily used with a Wi-Fi internet connection although the purchaser had the option of also connecting it to a cellular provider for cellular phone service if desired.

Apple iPad, second generation

Tablet computers are extremely light and mobile. They are great for watching videos, viewing photos, surfing the web, handling email, reading books, and producing documents on the run. On the other hand, laptops usually have larger displays, physical keyboards instead of virtual keyboards, and more powerful hardware and available software.

The tablet computers now come in all types of configurations. For example, there is the Microsoft Surface tablet which is attachable to a keyboard. Another tablet computer, the Lenovo Yoga, is a folding tablet in which one of the folded sides is a keyboard. These types of tablets are lighter than laptop computers, and they have a physical keyboard which a large segment of the population requires.

Then there is the Amazon Kindle, which was really the first tablet to revolutionize how books are read. The Kindle was introduced to the public around the same time as the iPhone. It's an e-reader in which books, newspapers, magazines, movies, and any other digital media can be downloaded rapidly via the Kindle store. Instead of carrying around heavy paper books, they can now be downloaded and easily read on this tablet which is extremely lightweight. Millions of people love carrying a number of books, newspapers and magazines on this lightweight device although other millions enjoy handling a physical book, newspaper or magazine. After its introduction, there was the Kindle Fire which was more versatile, in that it enabled web access and many of the other functions that computers have. There are numerous versions of the Kindle, even a Kindle Keyboard which includes a physical keyboard.

The iPhone has been discussed in an earlier chapter of this book concerning telephones because there's no doubt that the iPhone is a type of mobile phone. It has also been discussed in a later chapter of this book concerning digital cameras. But isn't the iPhone also a computer? If it had a physical keyboard and a much larger display it would have all of the characteristics of a computer. Let's think about that. It's easy to connect a physical keyboard to the iPhone using Bluetooth. Likewise, it's easy to connect a monitor to the iPhone using a USB to HDMI cable. If you want, a mouse could be paired to the monitor using Bluetooth. There you have it— a powerful Apple computer basically comprising an iPhone, millions of times more powerful than the giant computers that I programmed in college.

PROTECTING COMPUTER PROGRAMS --THE CHESS COMPUTER CASE

In the 1970s, with microprocessor-circuits proliferating, the world was now changing over to computer-controlled electronic devices. The programs for these devices were typically embodied on ROMs which could easily be copied. A ROM is a *read-only memory* device that stores computer-readable code. How could these programs be protected? There were no legal cases providing an answer until the chess computer case, *Data Cash v. JS&A*, in which I was the attorney for JS&A.

The technology involved was a hand-held chess computer introduced in 1977 by DataCash named "Compuchess." You would play chess against the computer using a separate chessboard. Chessboards have 64 squares with a conventional coordinate system, seven across (A to H) and eight up-and-down (1 to 8). If you wanted to move a piece from E2 to E4 you would enter "E2E4" on the computer keyboard and this would be shown on a display on the chess computer. The chess computer would show its move, for example, E7 to E5 as "E7E5." Prior to play, you could select one of seven levels of play, from "beginner" to "highly advanced."

Compuchess (left) and JS&A (right) hand-held chess computers

In 1979, JS&A begin marketing the JS&A hand-held chest computer. The chess computers were purchased from a Hong Kong manufacturer, Novag Industries. JS&A did not know at the time that those computers contained a program that was copied from DataCash's Compuchess program. Novag had duplicated Data Cash's ROM. The program was in the form of a computer-readable code embodied on the ROM.

DataCash brought a copyright infringement suit against JS&A in federal court in Chicago, requesting a preliminary injunction. As defendant JS&A's attorney, I studied the situation and determined that we had a viable defense. Under the United States copyright laws, every work published prior to 1978 had to contain a copyright notice. Without that copyright notice, the copyright owner lost its rights to a copyright. My study of the Compuchess chess computer revealed that it was published in 1977 with no copyright notice on anything; not even the on the ROM or the code embodied within the ROM. I moved for summary judgment on the ground that DataCash forfeited its copyright by its failure to provide a copyright notice.

The court issued a detailed decision in our favor. However, the basis of the decision was a surprise. The court held that computer programs were not copyrightable! This blockbuster decision shocked the entire computer industry (as well as me). As a result of the decision the US Copyright Office discontinued issuing copyright registrations for computer programs.

DataCash appealed to the US Court of Appeals. They argued that computer programs are copyrightable. Intel Corp., the world's largest chip manufacturer, also filed a brief arguing that computer programs are copyrightable. I filed a responsive brief, fully agreeing with DataCash and Intel that computer programs are copyrightable. But I also argued that DataCash forfeited its copyright by publishing in 1977 without a copyright notice. I asked the Court of Appeals to affirm on that other ground.

We won. The Court of Appeals held that DataCash forfeited its copyright by not having the required copyright notice. This decision implied that computer programs are copyrightable and is the landmark case on that issue. After that decision I met with the Registrar of Copyrights and other US Copyright officials and convinced them to grant copyright registrations on computer programs embodied on ROMs or any fixed storage device,

The computer program embodied on a storage device is in the form of a computer-readable code. For example, in the iPhone the operating system comprises a code stored in a storage device in the iPhone. This computer-readable code, like the words of a novel, is what is covered by the copyright registration. A computer program subject to copyright could be much simpler, such as the program for the operation of your home coffee machine.

Being able to register a copyright on computer programs was a blessing. Over the years I filed numerous copyright infringement lawsuits in federal court for clients whose programs were copied. An example of this involved an Italian electronic organ manufacturer that I represented, General Electro Music ("GEM"). GEM discovered at a trade show in Chicago that an Italian competitor, Viscount, was selling organs having ROMs which contained a copied program of GEM's organ sounds. Because of a pending lawsuit in Italy, GEM had previously "dumped" the competitor's ROMs and compared the programs, finding the codes to be identical. I obtained a copyright registration for GEM's program on an expedited basis and quickly prepared and filed lawsuit papers, requesting a temporary restraining order. The judge compared the identical programs and granted the order, the defendant Viscount was enjoined from selling the infringing organ, and the case was eventually settled.

Returning to the involved technology, the chess computer has evolved into many forms, including computer chess boards that automatically move the opponent's chess pieces, desktop computer software for playing opponents over the Internet, and chess game apps for your smartphone. The computer software has become so sophisticated that world chess champions have been beaten by the computer starting with the defeat of champion Garry Kasparov by IBM's extremely powerful computer Deep Blue in 1997.

CHAPTER 6

Film Photography

FILM CAMERAS I HAVE USED

One of my hobbies in elementary school and in high school was photography. Through the years, I have seen extraordinary changes in the fundamentals of photography, how the image is created and processed, how cameras are designed and how they operate, as well as how the public views the photographic image.

In the 1940s, my father would often carry his camera when we were outside or went on car trips, and he enjoyed taking snapshots of the family. Occasionally I would ask for the camera and I always enjoyed taking pictures. At that time and up until the late 1990s, there were no digital cameras. Instead, cameras used photographic film which most people born after 2000 have never seen.

Kodak 620 Film bellows folding camera, circa 1940

My father's camera was a Kodak folding camera using 620 film, identical to the camera shown above. The camera was very easy to use and my brother and I borrowed it from my dad occasionally, when we were in elementary school, before we had our own cameras. It only had two shutter speeds, "T" for a time exposure and "I" for a shutter exposure of about 1/100 of a second. The lens aperture opening was fixed so the camera really was designed for outdoor snapshots.

In the 1940s, Kodak was the largest camera manufacturer in the United States. Most of the black-and-white camera film was manufactured by Kodak and essentially all of the color film was manufactured by Kodak.

Most cameras used either 35mm film, 620 film, or 120 film. The numbers 620 and 120 were really like model numbers. The large studio and press cameras typically used sheet film such as 4 x 5 sheet film. The 35mm, 620 and 120 film came in rolls. The 35mm film was typically rolled up inside a cylindrical light-blocking cassette while the 620 and 120 film came in a roll with the film wound within light-blocking paper.

The film was a strip formed of plastic. One side of the film was coated with chemicals including a gelatin emulsion containing microscopic silver halide crystals. Since silver halide is light-sensitive, upon exposure to light it will form an image which can be developed with chemicals. The chemical processing of the film is discussed later in this chapter.

Although it was within a cassette, the 35mm film in the cassette was called a "roll" of 35mm film. It typically could include 20 or 36 images. Depending upon the camera, a 620 and 120 film typically could include between eight and 16 images. The paper in which the 620 and 120 film was wound had numbers on the outside, for example numbers 1 through 8. The numbers would align with a small window on the back of the camera. Each time you would take a picture, you would turn a knob on the side of the camera to advance the film to the next number. See the folding knob on the Kodak camera shown above.

Some cameras had a lever which would automatically move the film for the next exposure. All 35mm cameras had those levers with a number wheel indicating the number of times the lever had been advanced. Although using the lever was easier and quicker than winding the film to align the next number in the window, an unusual problem could occur. The following scenario happened to me a couple of times.

When I loaded a roll of 35mm film into my 35mm camera, I placed the cassette of film in a holder on one corner and I brought the beginning of the film around the backplate into a take-up spool on the other corner. I was not careful to be certain that the beginning of the film firmly engaged the take-up spool. Because the beginning of the film had not firmly engaged the take-up spool, the film released when I moved the lever to advance the film and I did not realize it! Then, each time I took a picture and advanced the lever, the number wheel moved to show the next number but in actuality the film was not moving. When I thought I had taken all 24 pictures on the roll I then "rewound" the film into the cassette. However, in reality the film had remained in the cassette all the while I was taking the pictures. I then sent the film to the lab for processing and it came back blank because it had never been exposed!

After film is exposed and processed, the images on the film are called "negatives" because they are negative images. The portions of both black-and-white film and color film that are exposed to more light are darker. With negative color film, the film colors are the complementary colors to the original color. For, example a blue shirt on a person would be yellow in a color negative. Later in this book there is a discussion of how photographic film is processed to obtain prints and enlargements.

A photo negative of me, at 13 years old, on a delivery bike

I obtained my first camera while I was in elementary school. It was an Argoflex Seventy-five, a twin lens reflex camera that was extremely popular at the time. It was very simple to use and it had an attached flash unit which used flashbulbs. You would buy a package of flashbulbs and each one would last for only one flash. The lens aperture was fixed, and it had only two shutter speeds: “Time” for a time exposure and “Inst” for 1/100 second. The flash unit used two D batteries and the reflector had a bayonet mount for the bulb.

Argoflex Seventy-five, with attached flash unit

When I was in the eighth grade, I decided that it would be useful to obtain a camera with more features to enable me to take more "professional" photographs. I obtained a used German Voightlander twin lens reflex. It had a variety of shutter speeds, variable lens aperture adjustments starting with f/4.5, and a fine multi-element glass lens. With this camera, I was even able to take pictures indoors as long as there was a substantial amount of light, without a flash.

The difficulty of taking pictures indoors without a flash was that photographic film at that time was relatively slow. The main black-and-white films were Kodak Plus X and Kodak Super XX. Super XX was twice as fast as Plus X but it also resulted in a grainier picture. When I was in high school, Kodak introduced Tri X film which was even faster and the negatives could be enhanced by using extra development time. However, faster film generally resulted in grainer negatives.

Early in high school I was able to obtain a very unusual antique used camera called the National Graflex II. This camera was strange looking and had some very special features. It was a single lens reflex using 120 size film, providing 6 cm X 6 cm negative images. It had numerous shutter and lens aperture adjustments. It used a focal plane shutter, instead of a leaf type shutter, just like the focal plane shutters used by most of the 35mm cameras. It had a Tessar lens, which was a very fine lens. It was like a giant 35mm SLR camera.

National Graflex Series II Film camera with Tessar Lens
Photo credit: Cory Hagelstein

It was so interesting using the Graflex camera. Looking down into the viewfinder you would see the image projected from the single lens of the camera. An angled mirror was used to project the image into the viewfinder. As soon as you pushed the shutter button, the mirror would flip up with a loud noise and the image would be projected onto the film. The mirror would stay up after the picture was taken and the viewfinder showed a blank image. Also, the shutter made a loud whizzing sound when it was activated. The focal plane shutter comprised a motorized curtain with a vertical slit through which the image was projected. The speed of the slit traveling across the image was the length of the exposure. This camera was a real conversation piece!

The noise created by the camera when I took a picture created a problem once. My brother and I went to a burlesque show at the Hudson Theatre in Union City, New Jersey and I brought the camera with me. We sat in the front of the balcony and I began taking pictures during the show. Although I tried to hide the camera while I was taking pictures, an usher heard the noise from the camera and he confiscated the film. However, we brought a different camera at a later time and took some interesting photographs.

In 1954 I decided to purchase a professional camera since I had become extremely interested and proficient in photography. The camera that I purchased was a Rolleiflex, a German twin lens reflex camera that was used by professional photographers around the world.

Rolleiflex twin lens reflex camera,
Zeiss Tessar f/3.5 75 mm lens, 1954

I enjoyed using my Rolleiflex for many years. It was a beautifully made, well designed camera. The viewfinder presented an excellent image and it had a wide range of variable shutter speeds and lens aperture openings. It had a focus knob on the left side and a crank on the right side (as shown in the photograph), for moving the film to the next picture. The camera used 120 size film and was set for twelve 6 cm x 6 cm pictures per roll of film.

The Rolleiflex had MX flash synchronization which allowed you to plug in an electronic flash unit in sync with the camera shutter. To use the electronic flash sync, you would set the shutter for 1/125 second or less. The flash, having a duration of 1/2000 second or less, would fire while the shutter was open. The operation of electronic flash units is discussed below in another chapter of this book. The Tessar lens was wonderful.

When I took the subway from my Queens apartment to Manhattan, I would usually carry my Rolleiflex with me. One of the photos that I took in Manhattan, which I called "City Bird," won an award from *Popular Photography* magazine in 1955.

City Bird, New York City, 1954

In the late 1970s, after I started using my Nikon as discussed below, I gave my Rolleiflex to my sister. She was very attached to it and thought it was a beautiful piece of machinery. She had gotten into photography because of the interest that my brother and I showed as well as having that camera. When she lived in Mexico (for three months) more than 40 years ago, she used the Rolleiflex constantly. She still keeps on her desk one of the photographs she developed herself taken with the Rolleiflex.

One feature that was missing from this camera was interchangeable lenses. In 1974, I purchased a Nikon F2 Photomic camera, which was a 35mm camera with interchangeable lenses. The basic lens was a 50 mm lens, but I also had a 28 mm wide angle lens, a 90 mm portrait lens and a 200 mm telephoto lens.

Nikon F2 Photomic 35 mm single lens reflex camera

My Nikon camera was made in Japan and was a very solid, professional, highly precision-made camera. It had all of the "bells and whistles" and the lenses were easily interchangeable. For its size, the camera and lenses were quite heavy. Nikon lenses, which were named Nikkor lenses, were considered some of the finest in the world. I greatly enjoyed using my Nikon camera for many years.

However, I found that traveling with the Nikon camera was difficult at times because of the bulk and weight of the camera and lenses. I decided to purchase a camera that was much more portable. After substantial research, I chose what I believed to be the smallest, precision made, finest 35mm camera on the market. It was the Rollei

35, invented and made by the same company that made my Rolleiflex twin lens camera.

Rollei 35 camera with F/3.5 40mm Tessar lens
Photo credit: top_camera_japan

The Rollei 35 was a viewfinder type camera and everyone who saw it marveled at how small a full-function 35 mm camera could be. The camera was only 3 ¾ inches wide, 2 ¼ inches high, and 1 ¼ inches deep. It weighed less than 12 ounces. It had an F/3.5 40 mm Zeiss Tessar lens and all of the appropriate shutter, film and focus adjustments. It had a built-in battery-operated light meter. Its small size enabled it to be fit into practically any pocket. In order for a 35mm camera to be so small, the cocking lever and rewind crank were located in unusual places on the camera, and the film compartment was unique. Although it didn't have interchangeable lenses, it was a pleasure to travel and take pictures with this little jewel.

All of the cameras discussed so far were film cameras which required that the exposed film be processed by a film processing service or in one's own darkroom. This prevented you from seeing your picture for a substantial amount of time after the picture was taken. The only kind of film camera that enabled you to see the picture within a very short time was the Polaroid camera, introduced in 1948.

The Polaroid camera enabled you to take a picture, process the Polaroid film within the camera, and the resulting photograph was on paper in minutes. I remember seeing Polaroid cameras when I was in elementary school and although I was fascinated by the technology, I didn't like the concept of having only one photograph without a negative, thereby preventing the printing of multiple prints.

The first Poloroid camera, Model 95, 1948

Polaroid sold film packs that you would insert into the Polaroid camera. After taking each picture, you would hold on to one of the tabs, pull out the exposed film, wait one minute and then peel off the photographic paper from the remaining product. The photograph would be in a state of partial development and after another few minutes it would be a complete photograph on the paper. However, the developing chemical residue would often come off on your hand. The quality of the photograph would usually not be as good as the quality of a photograph from a fine film camera but at least you were able to see the results within a couple of minutes instead of having to wait for several days.

During the 1970s Polaroid introduced a model, the SX-70, which ejected the film automatically after the picture was taken. You could then watch the picture develop in your hand. It would take about five minutes. I purchased one and enjoyed using it as an "instant camera." The SX-70 had many adjustments and the resulting photograph was not bad.

Polaroid Model SX-70 instant camera

The Disposable Camera

Although film cameras have generally been displaced by digital cameras, there is one type of film camera that has lasted since the mid-1980s. That is the disposable camera, which is extremely inexpensive and easy to use.

Fujifilm 35mm disposable camera

The disposable camera is intended for single use, meaning it is meant to be used with a single roll of film which is pre-loaded inside the camera when the camera is purchased. The typical disposable camera costs only about $10 and it includes a built-in viewfinder, electronic flash, fixed-focus lens and shutter system, an AA battery, and a film advance mechanism.

The camera also includes a built-in capacitor charging circuit for providing a high-voltage to the electronic flash. As discussed in the next chapter, it is the high-voltage to the flash tube that causes the xenon gas within the flash tube to break down and emit a brilliant flash.

Most disposable cameras use 35mm film which will enable taking 27 pictures. When the 27 pictures are taken, the entire camera is returned to a processing lab where are the pictures are developed. In the 1980s and 1990s, the photographer typically received prints of a designated size, for example 4 inches x 6 inches. More recently, the processing lab emails digital images to the photographer or provides a digital disc.

Taking photographs with a disposable camera is really simple. There is no ability to focus or to change the lens aperture. You simply advance the film and point and shoot. If you want to use the electronic flash, you slide a tab in advance of taking the picture. A red light pops up to indicate that the capacitor is charging to a high-voltage. When you shoot, the shutter will automatically be synchronized with the flash.

I have no idea how many times one camera is recycled, but I am aware that there are machines used for the recycling of disposable cameras. Almost all of the parts can be reused and this reuse aids in maintaining the low price of the disposable camera.

Before camera phones were introduced, I remember attending many wedding receptions, bar mitzvah receptions, and similar occasions at which disposable cameras were provided on the tables for each guest or couple. We would take pictures of the festivities and we could then bring the cameras home to have the film processed.

How We Processed Film

With the exception of instant cameras such as the Polaroid camera, photographs taken by film cameras required separate processing, usually referred to as developing and printing. The image could not be viewed until the film was processed (this was one of the basic differences between film photography and digital photography). You could bring or mail the exposed film to a film processing laboratory or store which accommodated film processing, or you could process it in your own darkroom if you were so inclined.

I used all types of photo processing. Some of my films were dropped off at a drugstore, some at a camera store, some were mailed to a photo processing laboratory, some were mailed directly to Kodak for color processing, and many films I developed and printed myself or with my brother in a darkroom.

Until around 2010, a number of the larger drug store chains did film processing. At first, they would send the film to an outside processing plant and it would take a few days before the processed negatives and prints would be returned. Likewise, you could drop off your films at camera stores which would also send the films to a processing plant. In the late 1980s, film processing machines became readily available to the drugstores and camera stores and they would process the films at the store, usually within a couple of hours from when they were dropped off. This made it possible to see your prints and enlargements of photographs that were taken just a few hours earlier.

In 1957 I had a summer job with Peerless Camera store on Lexington Avenue and 43rd St. in Manhattan. Although part of the time that I was there I sold movie cameras, I spent most of my time in the film processing department. Peerless was one of the largest camera stores in the country at the time. My department had four employees and we sent the customers' films to an outside processing lab. The lab handled developing, printing and enlarging of every size and type of photographic film. Processing would usually take from two days to a week.

Occasionally the lab would send a customer's negatives back without any prints. A person at the lab had determined that at least one of the negatives showed a person not dressed "appropriately." This was the 1950s. Even a photograph a of a baby in a bathtub was inappropriate.

Because Peerless Camera was in midtown Manhattan there was heavy traffic in our department and often a celebrity would drop film off. I remember when Walter Cronkite, the most famous news anchor at the time, dropped off about 100 rolls of 35mm film from a European trip that he had taken.

There were numerous mail-order film processing labs, some of which were very popular as a result of their low cost and some provided extremely high custom quality. But when I was in elementary and high school, the color film that I used had to be sent to Kodak for processing. It was Kodacolor color negative film, which practically everyone used for color prints. Kodak had a secret process for developing the Kodacolor film and it came with free developing using a Kodak pre-paid mailer.

Kodak Kodacolor II 35mm film
with mailer (attached at rear)
Photo credit Craig M. Eisenberg

In the mid 50s the Department of Justice charged Kodak with antitrust violations and Kodak made the process available to the industry. I always felt that the antitrust case hurt the public more than it helped. Kodak lowered its price on Kodacolor film very slightly. Other labs using the development process charged separately, with the total charge for developing Kodacolor film becoming greater than it used to be when Kodak bundled the development with the sale of its film.

Using A Darkroom For Processing

To really get immersed in photography, it was best to process black and white films in your own darkroom. My brother and I used a bedroom in our apartment and the adjacent bathroom as a darkroom during elementary school and in high school, and we enjoyed developing, printing and enlarging our own black and white films.

The first thing was to remove the film from the camera and develop it to obtain negatives. To that end, we used a film developing tank which had to be loaded with film in absolute darkness. Since the film was sensitive to light prior to its development into negatives, any light impinging on the film would expose the film. The developing tank basically included a spiral reel for receiving the film, a liquid container for receiving the spiral reel, and a top closure for the container. The top closure had a removable cap for receiving the chemical solutions.

A stainless steel film developing tank

To load the developing tank with film, you would bring the film container and the developing tank to a room that was absolutely dark. In the dark, you would have to use feel instead of vision. You would unwind the film from its container and insert it into the spiral reel carefully so that the end of the film followed the spiral into its center. You had to make sure that no part of the film was bent or was touching another part of the film. You then placed the spiral reel containing the film into the container and closed the top. Once the developing tank was closed, no light was able to enter the tank and you could now turn on a light and add the chemicals to develop the film.

The developing chemicals would be added through the top opening, by removing the top cap from the container closure. The closure was constructed so that light could not enter the container, even when the top cap was removed. First, liquid developer would be introduced into the container. The developer used was dependent upon the type of film and other criteria. After a certain amount of time, the developer was removed and a stop bath solution, typically acetic acid, was introduced. This stopped any further development of the film. The stop bath solution was removed and then a fixer solution was introduced. The fixer was removed and then the container and reel were washed in water to remove all chemicals. The film was then hung to dry and development was now completed.

Then it was time to either print or enlarge the pictures. I had both an Airequipt Junior contact printer and an enlarger in my darkroom. When we were young, my brother and I shared a bedroom and we kept our darkroom equipment on top of a dresser. To use the bedroom as a darkroom we would close the shades and blinds and use only a red or amber darkroom light. The photographic paper which would be exposed to light to obtain the photograph was not sensitive to red or amber light and so it was unnecessary to do everything in absolute darkness. However, since the photographic paper was sensitive to other than red or amber light, it had to be handled carefully and opened only in the darkroom light.

If you wanted prints the same size as the negatives, a contact printer was used. If you wanted larger prints, an enlarger was used. I used four trays for the necessary chemical solutions. Similar to film development, the first tray contained developer, the second tray contained a stop bath solution, usually acetic acid, the third tray contained acid fixer, and the fourth tray contained water. After the photographic paper was exposed to light passing through the negative, it would be placed in the developer tray and you would watch the photograph develop. When it appeared satisfactory, you would place it in the stop bath, thereafter in the fixer tray and then in the water tray.

Airequipt "Junior" contact printer, circa 1952
Photo credit: Bill Sanders

An enlargement used the same chemical processing as the contact printer but the photographic paper was exposed to light from the image being projected from the enlarger. To this end, the negative was inserted into the enlarger and the enlarger was moved to the desired position and focused to project the desired image upon the photographic paper. After this exposure, the exposed paper would be placed in the developer tray to start the developing process. Most of my enlargements were 8 X 10 inches so I used trays that handled at least that size paper.

A vintage Leica Focomat enlarger with an easel

The photograph *City Bird* shown earlier in this book is an enlargement from a roll of film I developed and then enlarged from my darkroom enlarger on 8 X 10 inch Kodak photographic paper in 1954.

CHAPTER 7

Digital Photography

The difference between film cameras and digital cameras is tremendous. As discussed in the previous chapter, a film camera uses chemicals on film to capture the photographed image and then liquid chemicals to develop the film. The images are stored on the developed film (for example, in the form of negatives), for printing or enlarging using additional liquid chemicals. By contrast, a digital camera uses electronic photo sensing devices to capture the photographed image. The captured images are then electronically digitized and stored as a digital file on a memory device. The stored images can be digitally processed, viewed, printed, emailed, etc. With digital cameras, the darkroom is a thing of the past.

If you were born after the year 2000, you probably have never used a film camera. Every camera that you have ever held or used is probably either a dedicated digital camera or a cell phone incorporating a digital camera. I had always enjoyed using film cameras and never heard of a digital camera until the 1990s. Occasionally I would read articles concerning the advent of digital cameras but they were unavailable in the camera stores.

Until the mid 1990s, I did not find that the camera stores or electronic stores were selling digital cameras. Then I started noticing advertisements and articles about digital cameras that really intrigued me. A client of mine in the mid 1990s had a very expensive Nikon single lens reflex camera with a Kodak digital camera back. He used it in his advertising business. I found the quality of the photographs to be excellent and noted the advantages of being able to take hundreds of photographs, see them immediately, store them immediately, and be able to use these high-quality photographs as soon as they were taken without any kind of chemical processing required.

Although professional digital cameras were becoming available in the late 1990s, they were very expensive and most professional photographers, who were expert with film cameras, did not want to use them. However, the consumer market became wide-open and by about 2000, digital cameras were much lower in price and were selling by the multimillions.

I Start Using A Digital Camera

I purchased my first digital camera in the late 1990s. Although I planned to continue to use my Nikon F2 film camera for taking photographs around my home, I thought it would be useful on vacations to have a camera that was small, lightweight, fit in my pocket, and produced immediate photographs. My first digital camera was a simple "point and shoot" camera, a Canon PowerShot.

Canon PowerShot A720 digital camera

I was thrilled with the quality of the digital photographs, the very small size and weight of the camera, and how easy it was to take satisfactory pictures. You could select the resolution of the photograph, depending upon the amount of storage you wanted to use. The camera even had a small built-in electronic flash which actually provided a substantial amount of light. The camera used a memory card that could store up to 2GB, far less than the memory cards today.

A problem that was concomitant with the early digital cameras was the time leg between the actuation of the shutter and the capturing of the image. You would press the camera shutter button and the picture would not be captured immediately. The time lag on some early digital cameras was as much as ½ second. Unfortunately, some good pictures were missed. However, this time lag problem was corrected as the industry progressed.

After using the Canon PowerShot for several years, I acquired another very small "point-and-shoot" type digital camera, a Nikon Coolpix. I kept it at my office and used it often in my business. The camera weighed only about one pound, was as small as the Canon, and had higher resolution then the Canon. In addition, I was able to take video with the Nikon.

Digital cameras were now in wide use, far outnumbering the use of film cameras which were becoming obsolete. Even the top professionals were giving into the digital revolution, discontinuing the use of film cameras in favor of high quality professional digital cameras. But what happened to Kodak? It's really a story of bad decisions made by a company that at one time was a corporate giant. Kodak actually invented the digital camera in the 1970s. But when digital cameras started to become popular around 2000, Kodak decided to hold back because it thought digital cameras could ruin its film business. The Japanese companies, such as Nikon, Canon, Sony, Panasonic, Olympus and Fuji recognized that the market was really in digital cameras, not photographic film. Kodak should have tried to sell more cameras, not more film. In 2012, Kodak filed for Chapter 11 bankruptcy and later announced that it would discontinue selling digital cameras.

At my 70th birthday party in 2009, my children gave me a Sony Cyber-shot DSC-HX1 digital camera that I still have. This camera was really ahead of its time and it incorporated all of the features that a non-professional photographer would want to use.

Sony DSC-HX1 digital camera, with flash in up position and lens partially extended

It has a 20X zoom lens which was phenomenal at that time. The lens is optically equivalent to a f/2.8 to f/5.2, 28mm to 560mm lens. By moving a lever adjacent to the shutter button, you can go from wide angle to super-telephoto. It uses a 10 megapixel CMOS image sensor. The camera weighs only slightly more than one pound and is comfortable to hold.

The camera is replete with adjustment dials and buttons. Among many other things, you can adjust shutter speed, aperture, ISO sensitivity, flash options, display options, white balance, metering, color, face detection, contrast, sharpness, and shooting modes. The shooting modes include manual, semi-manual, intelligent-auto, panorama and video. It is an excellent video camera, with high definition up to 1080p resolution and the ability to use the 20X zoom lens during the taking of a video.

The LCD display on the rear of the camera, shown below, can be tilted, is 3 inches in diameter, and can be filled with all sorts of information using the various buttons. It toggles with an electronic viewfinder that is useful for fast moving scenes. The pictures are stored in a memory card that can be purchased with a capacity up to 32GB. The camera also has some fixed, internal memory.

Rear view of Sony DSC-HX1 digital camera

Today the semi-professional and professional digital cameras seem to fall into two categories: the DSLR (digital single lens reflex) camera and the mirrorless interchangeable-lens camera. One of the main differences between these cameras and the more casual digital cameras such as those discussed above is the size of the image sensor. Typically, the image sensor used in semi-professional and professional digital cameras is significantly larger than the image sensor used in the more casual digital cameras. The other main difference is that the professional digital cameras use interchangeable lenses instead of fixed, zoom lenses.

The DSLR camera is probably the type most used by professionals. It is relatively heavy and bulky, partially due to the use of a large image sensor and the internal mirror system. The mirror directs the image from the lens to the optical viewfinder prior to actuation of the shutter. When the shutter is actuated, the mirror swings up and the image is directed from the lens to the image sensor. In this manner, prior to taking the pictures, the photographer has an accurate optical preview of the image in the viewfinder. This is different from seeing the image on an LCD screen as with my Sony digital camera.

However, most of the more recent DSLR cameras incorporate an LCD or OLED display screen on the rear of the camera and enable what is referred to as a "live view" on the display screen. To accomplish this, if the photographer wants to use the "live view," a switch can be actuated to toggle the mirror to its upward, locked position. In this manner, the image from the lens is directed on the image sensor instead of being directed into the viewfinder, prior to actuation of the shutter. This "live view" is particularly appropriate when the DSLR has video capability. Many of the present DSLRs are top quality video cameras as well as being top quality still cameras.

The term "mirrorless" camera typically refers to digital cameras that do not use a mirror but instead the images are directed from the lens to the image sensor at all times. My Sony camera that is shown and described above is an example of a mirrorless camera. The term is also applied to cameras which have a larger image sensor and use interchangeable lenses. Because there is no mirror, the cameras can be smaller and lighter than DSLR cameras, and thus are easier to carry. Some mirrorless cameras have a viewfinder in addition to the display screen. However, these viewfinders are electronic, not optical.

The Image Sensor

There is been a lot of discussion about the "image sensor." The image sensor in a digital camera is equivalent to the film in a film camera. The photographed image impinges upon the image sensor, which captures the light and the color of the image. The amount of light and color are converted into electrical signals by the image sensor. The electrical signals are digitized to provide the information for forming the image on the display and for feeding to the memory for storage. It is all handled electronically; no chemicals are involved as in film cameras.

There are two basic types of image sensors, the CCD and the CMOS. The CMOS image sensors are widely used in digital cameras. These sensors are based on MOSFET technology which is described in the section concerning transistors in a previous chapter. The CMOS image sensor contains millions of transistors corresponding to the number of pixels that it will handle. We all have heard the word "pixels" being used in connection with the resolution of an electronic image. The greater the number of pixels, the greater the amount of detail in the image. If an image was 3000 pixels wide and 2000 pixels high, it would be a six megapixel image. An image sensor would be required to have at least 6 million transistors to handle the six megapixel resolution.

The image is composed by these pixels, each of which is a separate "dot." If the resolution is too low, or if the image is greatly enlarged, it will become "pixelated." Sometimes it can actually be interesting to see the pixels in the picture as illustrated in the enlarged photograph of lips that I took, below. As you move away from the picture, it becomes more realistic.

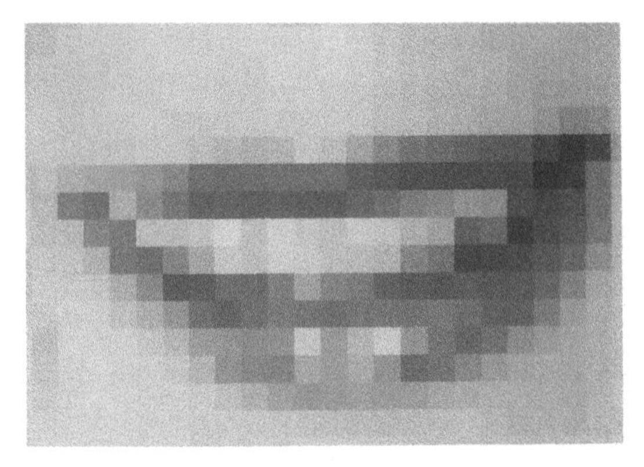

Digitally enlarged digital photograph showing pixilation

The Camera-Cellphone

Although CCD image sensors were used first, the introduction of CMOS Image sensors reduced the cost of image sensors, enabling the application of digital cameras to cell phones. The first commercial cell phone having a built-in camera was introduced in 2002. It was the Nokia 7650. It had a 0.3 megapixel rear camera and a 2 inch color display. The cell phone enabled the sending of photos over the cellular network and it had Bluetooth capability. For the next five years, most of the cell phones included built-in cameras but the displays were extremely small, usually about 2 inches in diameter.

The first commercial cell phone with built-in camera, the Nokia 7650

In 2007 the iPhone was introduced, and this changed the manner in which photographs and video were taken and displayed. Prior to the iPhone, the photographs and video taken with a cell phone were of very low quality even compared to an inexpensive point-and-shoot camera. Although the iPhone did not have video recording until the 2009 model, the first-generation iPhone introduced in 2007 had a nice two megapixel rear digital camera. There was no optical zoom, flash, or auto focus at the time. Further, there was no front camera for selfies. However, the photographs taken by the rear camera could be seen on the 3 ½ inch display screen, which was considered very large in 2007. The feature of having most of the front of the iPhone displaying the photographs was novel and exciting.

The Iphone 11 Pro And 12 Pro Camera Systems

Every year the iPhones camera was upgraded, typically with better resolution and more still photography and video photography features. This was always one of the selling points to induce the purchase of a new iPhone. An example of how much the camera system of the iPhone has been upgraded can be seen by first referring to the iPhone 11 Pro.

I know how it is to carry three separate lenses with a camera. I did it many years ago with my Nikon F2 Photomic 35mm film camera with three heavy lenses: a wide-angle lens, a normal focal length lens, and a telephoto lens. Now the ultra-lightweight iPhone 11 Pro (it weighs less than seven ounces) has four built-in cameras. One of the cameras has an ultra-wide lens, another camera has a wide-angle lens, and another camera has a 2X telephoto lens, all on the rear of the iPhone as shown below. Each lens has a 2X optical zoom in each direction and a 10X digital zoom. Each camera has 12 megapixel resolution. There is also a 12 megapixel front camera for Facetime and selfies.

Rear view of the iPhone 11 Pro, with three rear cameras and an LED flash

The iPhone 11 Pro camera system is replete with amazing features. Each lens is a high quality five or six element lens. Even the telephoto lens has a large F/2.0 aperture. There is a portrait mode with depth control enabling you to adjustably blur the background of the picture. The camera enables six different types of portrait lighting. There is a mode for taking panoramic pictures, a night mode, red eye correction, photo geo-tagging, auto image stabilization, a burst mode, and many other features.

The camera is excellent for video recording in high definition, up to 4K. A feature that I particularly like is that when you are taking a still photo, if you want it to become a video scene just continue to press the shutter button and it will go into a video recording mode. During video recording, you can zoom in and out. There is slow motion video, time lapse video, continuous autofocus video, and you can take still photos while you are recording in 4K video. There are dozens of other useful features that I haven't mentioned. It is easy to see why most people are using their smartphone cameras instead of larger digital cameras for taking photos as well as for video recording. Although this discussion has been primarily directed to the iPhone 11 Pro, it should be understood that competitors also have smartphones with excellent camera systems.

Apple's iPhone 11 Pro has a magnificent display. It is a 5.8 inch OLED "Super Retina" Touch display. There is also an iPhone 11 ProMax that has a larger, 6.5 inch display. The displays have such high resolution that you will never have to worry about seeing any kind of pixilation.

The iPhone 11 Pro, like most of its predecessors, has a built-in electronic flash. It looks like a white dot adjacent the three rear lenses. That is a light emitting diode (LED) which is also used as a flashlight. While recording video, you can activate the LED in order to provide additional lighting.

The iPhone 12 Pro was introduced by Apple in mid-October 2020, and includes several camera enhancements. The iPhone 12 Pro runs on an A14 Bionic chip having 11.8 billion transistors. Among other things there is a LiDAR scanner which is said to improve the photographic quality of still pictures and video recording. Apple claims that LiDAR is particularly useful for augmented reality (AR) applications. With AR, computer generated images are interposed with real images. A short YouTube video, showing how LiDAR is used with AR, can be viewed online at www.gerstman.com/lidar.htm.

The primary lens in the iPhone 12, which is designated as "wide," has an F/1.6 aperture and is a seven element lens. The telephoto lens has a 2.5X optical zoom. The "ultra-wide" lens is a 13mm lens having a 120° field of view.

The advantages of digital cameras over film cameras are evident. As soon as the photograph is taken by the digital camera, the resulting image can be seen, manipulated, used, and stored. Film cameras require that the photographs first be developed before they can be viewed. The digital images, which can be viewed immediately, can be deleted if the image is not satisfactory. It doesn't cost anything to take the digital photographs. The photographer can feel free to explore the scene and take as many photos as desired, in contrast to having to constantly purchase film and being limited to a small number of images allowed per roll of film.

In addition, digital cameras can store a huge number of images on their internal memory and on memory cards. Memory cards can store thousands of photos as well as storing videos. These photos and videos can be viewed anytime. Further, the digital images can easily be manipulated in ways that could never be imagined with respect to images on film. For example, the retouching of negatives from a film camera is cumbersome and complex, while the digital image can be "retouched" on the spot, even from the camera itself. Of course, software like Adobe Photoshop and numerous others enable an unbelievable amount of digital photograph manipulation.

Are there any advantages of film cameras over digital cameras? The only one that I can think of is that since film cameras are mechanical, they can be operated without requiring any electrical source. Digital cameras require electricity, which means that you must have the appropriate batteries available or the digital camera will not operate.

CHAPTER 8

The Electronic Flash

Essentially all modern digital cameras and smartphones include built-in electronic flash units. The miniature size of electronic flash units today amazes me. I have seen them evolve from a large, heavy device into a little dot!

USING FLASHBULBS

Looking back to when I was in elementary school, there were only film cameras. Flash guns for the cameras were typically attached to the camera and used disposable flashbulbs that could only be used once. Each time you wanted to take a picture using the flash, you would insert a new flashbulb into the flash mount that was in a strategic location on the reflector of the flash gun. Flashbulbs were available for purchase at all camera stores and many general stores.

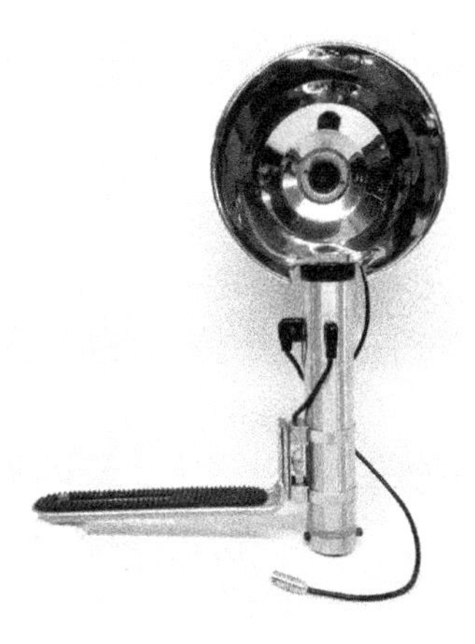

Vintage Trusite flash gun and camera bracket
with lower mount for screw-in base bulb
and center mount for bayonet base bulb

For indoor pictures, flashbulbs were almost a necessity because at that time the film was "slow" and it was difficult to take photographs using only available light. Although it didn't matter for black-and-white films, when certain outdoor color films were used, it was necessary to use a blue flashbulb in order to obtain the appropriate color of light for the film exposure.

The flashbulbs became hot when they were fired so you had to be careful removing them from the reflector socket. Also, if the bulb was defective and some air entered the bulb, the bulb might actually explode. It was wise to have a plastic shield in front of the reflector if close up photographs of a person were to be taken. Some bulb manufacturers such as Sylvania placed a blue cobalt spot on the bulb. The spot would turn pink if the bulb was defective and you would know not to use it.

Two boxes of Sylvania Press 25 "Blue Dot" flashbulbs

Electronic Flash Is Here

In the late 1940s, the electronic flash unit was introduced. Instead of one-use disposable flashbulbs, it used a permanent tube containing a gas such as xenon. The tube provided a high-intensity flash when the xenon within the tube was electronically ignited. To ignite the xenon gas, when the camera shutter was triggered an electrical circuit would rapidly discharge a high voltage from a capacitor, through the tube containing the gas. That became the principle of operation of the electronic flash unit. One flash tube could last for over 10,000 flashes.

The flash from an electronic flash unit was powerful and instantaneous. In effect, it substituted for the shutter of the camera. The camera shutter opened, the flash ignited for only a small fraction of a second, for example 1/2000 of a second, and then the shutter closed. The image on the film resulted from the high intensity light emitted by the flash. This instantaneous amount of light would freeze the action. The amount of light that was produced in that short time was enormous. The electronic flash unit was often referred to as a "speed light."

Triumph 1100-R6 electronic speed light
(From Triumph advertisement in July 1947 issue of *Popular Photography*, p93)

About the time that I started high school in 1952, I would occasionally see photographs in magazines and newspapers of press photographers with their Speed Graphic cameras and with electronic flash units. At the time, these units were very large, very heavy and very expensive. They required a separate power pack that was carried over the shoulder. I had no intention of purchasing one.

My brother and I were very interested in photography at the time, and we would often ride the subway downtown and look around the camera stores for bargains. New York had great camera stores, like Peerless Camera, Willoughby's Camera, and Olden Camera. We would purchase used and new darkroom equipment and things like lens filters, a camera case, film, flashbulbs, etc.

One day when we were at Peerless Camera we noticed a sign stating that when they opened Saturday morning, they would be selling a number of used cameras and related camera equipment at extremely low prices, to the first 25 customers. Peerless named it "The Peerless Gold Rush." Examples were given of $200 cameras selling for $20. We decided to get there early and see what we could find.

The next Friday night we set our alarm for 2 AM Saturday, and we took the subway around 2:30 AM to Peerless Camera. We found that there were already a couple of people camped out near the entrance. By the time the store opened at 9 AM, there was a long line of customers behind us. Shortly before the store opened, an employee had given ticket numbers to at least the first 25 customers.

As I entered the store, I quickly looked at the bargain merchandise that was available for the Peerless Gold Rush. I immediately saw a large electronic flash unit that was available for $15. It was in good condition and I believed that it had originally sold for at least $200. I don't remember the brand but I do remember that it had an extremely long flashgun with a big reflector and it was connected via wire to a large power pack. The power pack seemed to weigh over 5 pounds and had a shoulder strap. It contained heavy electronic components and a removable plastic-enclosed wet cell storage battery about the size of two packs of cigarettes. There was also a power line for connecting the power pack to an AC receptacle. At the store they connected the flash unit to a camera and it appeared to work perfectly. I was thrilled to obtain this great bargain.

I really enjoyed using this electronic flash unit. Action pictures were never blurry because the duration of the flash was so short that it froze the action. It provided an unusual amount of light: even subjects at a significant distance became well lit. However, about a year later when I turned it on one day I received a terrible electrical shock. Somehow there was an electrical short and I was at the wrong place at the wrong time. I decided that I did not want to continue to use this electronic flash unit.

I spoke with someone at Peerless camera and he told me that they usually had a couple of electronic flash units for sale at the Saturday Gold Rush. The next Saturday, my brother and I awoke even earlier and were the first on line at Peerless. I found another, more modern, used electronic flash unit and I recall that I was able to purchase it for about $25. It was a Dormitzer DB-1 unit and it had a much more contemporary look. The power pack, which weighed about 4 pounds, contained a large 45 volt "B" type dry cell battery. This was the type of battery used in portable radios and could easily be purchased at a radio store.

My Dormitzer electronic flash was a pleasure to use. I had it for many years and brought it with me to the University of Illinois, where I obtained a Bachelor of Science degree in electrical engineering. Being an electrical engineering student as well as an electronic flash unit enthusiast, I wrote an article about electronic flash units. The first page of the article includes a photograph of my flash unit and the article discusses the technology of electronic flash units at the time. It was published in the February 1959 edition of *The Illinois Technograph*, and is entitled "The Electronic Flash Unit." A copy of the article can be viewed online at www.gerstman.com/flash.pdf.

Using Transistors In The Flash Unit

After graduating from the University of Illinois, I was employed as a patent examiner at the US Patent Office in Washington DC, while in the evenings I attended George Washington Law School. During my stint as a patent examiner, I had a very interesting encounter with electronic flash unit technology. It happens that that the arts that I handled as a patent examiner were voltage regulation circuits and capacitor charging and discharging. This is exactly what electronic flash unit circuitry is all about.

Transistor technology was getting started. Because of this new technology, a number of inventors around the world filed patent applications in the late 1950s for a capacitor charging circuit using a dual transistor configuration. They saw that the circuit was ideal for use in electronic flash units. It would enable the electronic flash unit to be much smaller, use less battery power and use lighter electronic circuitry.

While I was a patent examiner, there were 11 patent applications on my docket which claimed this same invention. At that time, the broadest patent on an invention would be awarded to the first inventor. To that end, a patent interference proceeding was initiated to determine who first invented this new circuit to be used in an electronic flash unit. Although the final decision would be made by the Board of Patent Interferences, I was the patent examiner who was assigned to handle preliminary matters relating to this interference. As a result of that, I learned a lot about this type of electronic flash circuitry as well as about interference proceedings.

After a long battle at the Patent Office, a patent application assigned to Honeywell was awarded priority, a victory for Honeywell. Honeywell became a major force in this new electronic flash unit technology. Honeywell already had a very well received electronic flash unit on the market. It was the Honeywell Futuramic Strobinar electronic flash. It used three C size batteries in the handle. In their advertisements for the product, they promoted a "dual transistor unit" and how it has "No power pack" to carry over the shoulder. This was the start of more modern electronic flash units, in which the battery and circuitry could be located in the handle of the flash unit.

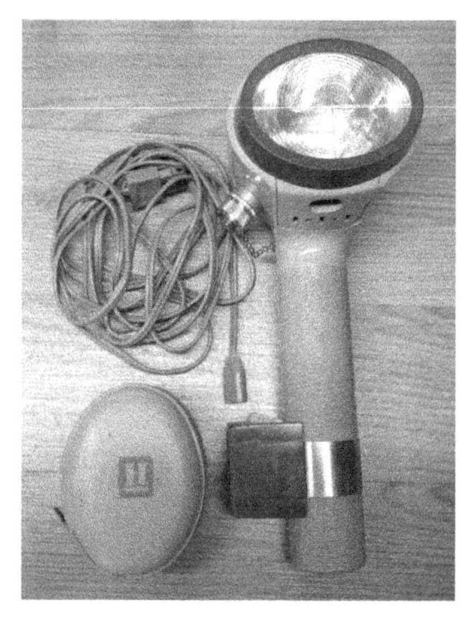

Honeywell Futuramic Strobinar flash unit

The Built-In Electronic Flash

Over the years, electronic flash units were shrinking in size and weight as a result of circuit miniaturization. In the late 1960s Voightlander introduced a 35mm rangefinder camera, named the Vitrona, that had a built-in electronic flash. The flash was not as powerful as the flash from earlier units, but it was a breakthrough. The size reduction from the past was amazing.

Voightlander Vitrona 35mm camera with integrated electronic flash

During the 1960s, I continued to use my Dormitzer DB-1 electronic flash unit, although I also had a much smaller flashbulb-type of flash gun that I sometimes used for portability.

In the 1970s, electronic flash units were finally becoming popular. Most of them connected to the accessory shoe or hot shoe on the camera. A hot shoe is a mount on top of a camera for attaching a flash unit. It provides flash synchronization.

The light intensity provided by these smaller units was not as great as the light intensity provided by the older, large units, but it certainly was sufficient. I purchased a Vivitar electronic flash which was very lightweight and small. It used an ordinary 9 V battery. It had a pivotable flash head so I could aim the flash upward and reflect the light off the ceiling instead of directly at the subject.

Vivitar Model 2000-D electronic flash
with hot shoe connector

I had a long connecting cord so I was able to hold the flash at a distance from the camera for optimum lighting. The long connecting cord that I used in the 1980s was invented by a client of mine. It was a hot shoe extension cord and it had a hot shoe at each end. One of the hot shoes connected directly to the camera's hot shoe. The other hot shoe was for holding a flash at a desired distance from the camera. You could also attach a flash to the first hot shoe. If you wanted to use more flash units you could connect more hot shoe extensions. I obtained US Patent No. 4,201,434 for my client on this very useful and successful invention.

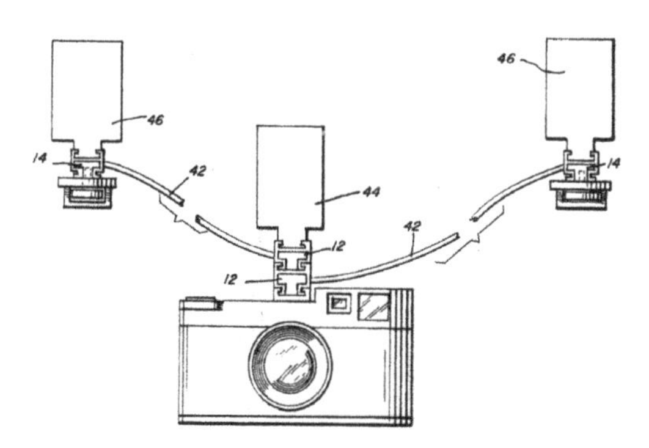

Figure 4 of US Patent No. 4,201,434

The electronic flash industry finally matured by the 1980s. Many cameras now had internal electronic flash units. One problem created by having the flash unit close to the lens was the creation of "red eye." It was not until later that red eye reduction technology appeared.

Similar To Implantable Defibrillator

In the 1980s, the implantable defibrillator was introduced to the field of cardiology. What could this possibly have to do with electronic flash units? The answer is that the circuitry used in the electronic flash unit is substantially the same circuitry that is used in implantable defibrillators. This came to my attention in the late 1980s when I represented a startup research and manufacturing company in the field of implantable defibrillators. I handled certain patent infringement lawsuits and patent applications for my client, and I became very familiar with implantable defibrillator technology.

Both implantable defibrillators and electronic flash units use a voltage regulated battery circuit in which the battery voltage is increased greatly to provide a high-voltage charge on a capacitor. The high-voltage charge that is built up on the capacitor is rapidly discharged when the capacitor is triggered. In a defibrillator, the capacitor is connected to the patient's heart. When triggered, the voltage from the capacitor gives the patient a high voltage defibrillating shock. In an electronic flash unit, the capacitor is connected to the flash tube containing xenon gas. When triggered by actuating the camera shutter, the high voltage from the capacitor ignites the gas in the flash tube to provide a burst of light. The similarity between the defibrillator circuit and the electronic flash unit circuit is interesting to say the least.

When digital cameras entered the photography scene, almost all of them, particularly the point and shoot type digital cameras, had a built-in electronic flash. The electronic flash was and still is based on the same principles as many decades ago but the size has been reduced greatly and this has reduced the resulting power of the flash. The digital cameras (except for smartphones) still use xenon gas flash tubes, particularly because the color of the light emitted by the xenon gas is similar to daylight and the flash is more intense than the flash provided by an LED light source.

Using The LED As Light Source

When cell phones entered the photography scene, there was not enough room for the conventional electronic flash unit using a gas flash tube. This problem created the start of using a light emitting diode (LED) as the source of the flash of light. The LED is a semiconductor device. When electrical current is passed through it, it emits light. DC current is supplied to the LED from the cellphone battery, and the LED is driven by an appropriate circuit between the battery and the LED. In order to provide a short flash of light, a short pulse of current is used.

The little white dot that you see near the camera lens on your smartphone is the LED. Most smartphones allow you to change modes with respect to the LED, including flashing for every picture, flashing only when a certain darkness level is sensed, and not flashing at all. You can also use the LED as a constant light source. So long as the appropriate electrical current is flowing through the LED, it will emit light.

The LED has many advantages over the gas tube as an electronic flash unit. For example, the LED is much smaller in size and weight. In fact, the LED is miniscule. The LED uses far less energy than the gas tube, so the LED gives the smartphone a longer battery life. The circuit driving the LED can be miniaturized to a greater extent than the circuit driving the gas tube. The duration of the light emitted from the LED can be adjustable from a short flash to continuous.

However, the LED has certain disadvantages over the gas tube. For example, the LED emits far less light than the light emitted by the gas tube. Some professional electronic flash units using xenon gas tubes can literally fill up a large room with light, even when the duration of the burst is very short. The color of the light emitted by the LED is not as close to daylight as the color emitted by a xenon gas tube. Some smartphones use more than one LED in order to obtain a greater amount of light, but even then, the amount of light is still much less than the amount of light that can be provided by a gas tube.

With film cameras, electronic flash units as well as flash units using flashbulbs were necessary in low light situations, in view of the relatively low sensitivity of the film, particularly color film. On the other hand, the image sensor in a digital camera and in a smartphone is so sensitive that excellent pictures can be taken indoors, even in a relatively low light situation. In the past, when I used a film camera, I typically would use a flash unit for indoor pictures. Now when I use my smartphone as well as when I use my Sony digital camera, I usually set the flash to the "off" position.

CHAPTER 9

Moving Pictures On Film

USING MOVIE FILM

I have seen an amazing evolution in the technology of taking and showing motion pictures on film in the past and now using digital video. Although I was very interested in photography when I was younger, I had no particular urge to take and show movies. Some relatives of mine had movie equipment, and it always seemed like a hassle to have to set up a movie screen, set up the movie projector, thread the film through the movie projector, and turn out the room lights in order to watch a short movie. However, when my daughter was born I felt it was time to start taking movies.

At the time, there were two different formats of movie film (other than the large 35mm film used with big studio cameras for theater releases by the major motion picture studios). The two formats were 16mm film and 8mm film. The standard 16mm film was 16 mm in width and had sprocket holes on both sides. The 16mm film for sound movies had sprocket holes on one side only and the other side was used for a sound track. The 16mm film was relatively expensive. A 50 foot length movie on 16mm film would run for only about 2 minutes.

16mm film clip, showing the sprocket holes

The 8mm film was an invention of Kodak in order to provide a less expensive alternative. The film itself had the same width as the 16mm film but double the number of sprocket holes. But taking movies required a number of steps. You would first insert a reel containing 25 feet of unexposed 8mm film into an 8mm camera. You would take the front of the film and extend it around a first sprocket wheel, then through a faceplate, then around a second sprocket wheel, and then insert it into a take-up reel in the camera.

During the taking of the movie, the 8mm movie camera would run the 25 foot reel of film through the camera exposing only one half of the film width. You would then open the camera, reverse the take-up reel with the original film reel, and then run the 25 feet through the camera again to expose the other half of the film width. Now you had 50 feet of exposed film. The exposed film was then sent to a processing lab, usually Kodak. The processing lab sliced the entire 25 foot length of film down the middle and spliced the ends together, resulting in a 50 foot length of movie film containing your movie.

Now you wanted to show your movie. This also required a number of steps. First you needed an 8mm movie projector to project the images on a movie screen.

Argus Model M-500 8mm movie projector

You would attach a take-up reel to a mount on one side of the projector and attach your movie film to a mount on another side. You would take a couple of feet of the front portion of the movie film and thread it through a first sprocket and rollers, then through a pressure pad, then through a second sprocket and rollers and sometimes through a third sprocket and rollers leading to the take-up reel. You would then insert the front of the movie film into a slot in the take-up reel. If everything was connected properly, you could then turn off the room lights, turn on the projector and project your movie onto a screen. Rewinding the film was easier because you didn't have to thread it through all those rollers and the pressure pad.

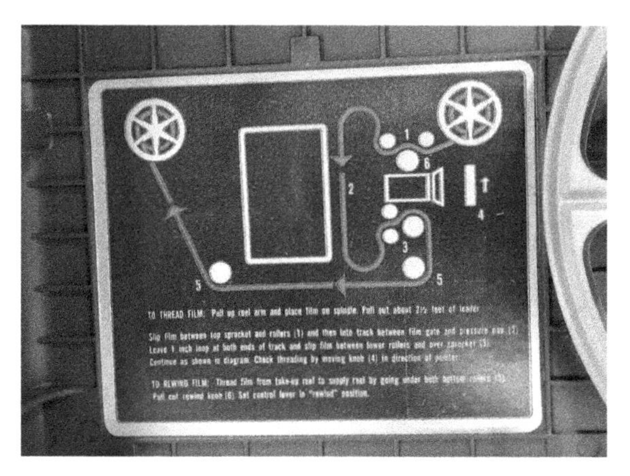

Threading and rewinding instructions,
located on the inside of the projector

The typical movie screen, also used with slide projectors, was foldable and portable. It had a tripod base and used a fabric having a slightly reflective surface. The fabric was spring wound into a hollow tube and you could pull it upward like a spring wound window shade.

Da-Lite Flyer collapsible projection screen
Photo credit: metoe83 (eBay)

The Electric Eye

Before I purchased a movie camera, I had a job selling them. During a summer break while I was in college, I was employed at Peerless Camera in Manhattan. Although my primary duties were at the film processing area, I was also assigned to an area where Peerless sold movie cameras.

The movie camera that was most popular at the time had just been introduced. It was the Bell & Howell Electric Eye 8mm movie camera having a novel feature. Other movie cameras required that you manually adjust the lens aperture according to the amount of light on the subject. This was often difficult if the amount of light was changing significantly while you were filming your movie. The

Bell & Howell Electric Eye camera sensed the amount of light on the subject and automatically adjusted the camera aperture.

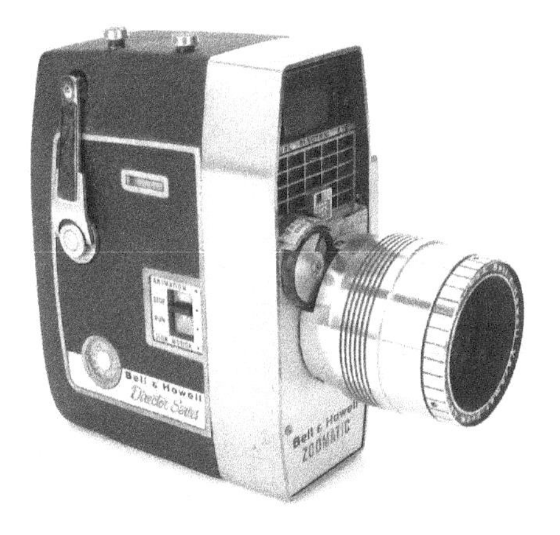

Bell & Howell Zoomatic Electric Eye
8mm movie camera

I never saw a movie camera that was so popular. Customers lined up to buy this Bell & Howell Electric Eye movie camera. At the time, Bell & Howell was a major force in movie projectors and movie cameras and this camera increased their prestige greatly.

My One And Only Movie Camera

In 1962 I purchased my first and only movie camera. From my experience selling movie cameras at Peerless Camera, I had learned that there was a film format designated 8mm magazine, that was much easier to use than the standard 8mm film. The film was wound inside a metal cassette which was called a "magazine." Instead of having to thread standard 8mm film through a standard 8 mm camera, you only had to insert the magazine into the camera.

The magazine was labeled "1st side" on one side and "2d side" on the other side. It held 25 feet of film. You would insert the first side into the camera, expose the 25 feet, and then turn over the magazine to the second side and expose the remaining 25 feet. Then you would send the magazine to the processing lab. They would slice the film down the middle, splice the two ends of the film together and return a 50 foot length of 8 mm film to you.

Many 8mm magazine cameras were manufactured and these cameras were dedicated to handling 8mm magazines. This seemed much simpler than having to thread the film through a camera so I began looking for an 8mm magazine camera. One Saturday while I was in Manhattan, I stopped at one of my favorite camera stores, Willoughby's, located on 32nd St. near Sixth Avenue. They had a large selection of used movie cameras and I found an 8mm magazine camera in excellent condition that seemed just right. It was a Revere model B-63. It was sturdy, very compact, and extremely easy-to-use. I was able to purchase it for $25.

Revere 8mm magazine Model B-63 movie camera

The camera had a standard lens. It also had a turret, as shown above, adapted to hold two additional lenses such as a telephoto lens and a wide-angle lens. It was easy to open the camera to insert or retrieve a film magazine.

The inside of the Revere 8mm camera with
an 8mm magazine for insertion

Taking Movies

At the time I purchased the Revere movie camera, I was working in Washington DC. When I returned to Washington, I took my first movie. I went to the Washington DC zoo with my wife and daughter and took the movie pictures very carefully. The camera was not an "Electric Eye" type so I had to adjust the lens aperture myself. My background in still photography greatly helped me in this regard. I used Kodachrome 8mm magazine movie film and sent the magazine in a mailer to Kodak for processing. It was returned to me, and I was thrilled with the results. Although it is almost 60 years later, I still have that first movie which has now been transferred from the 8mm film to a DVD.

My office at the United States Patent Office was in the Department of Commerce building in downtown Washington DC. Occasionally, when a foreign dignitary came to visit, President John F. Kennedy would have a parade and motorcade circle the sprawling Department of Commerce building which was and still is located between 14th St. and 15th St. and between Constitution Avenue and Pennsylvania Avenue. In order to promote these parades with the largest crowd possible, they would always take place between noon and 1 PM. I decided that these parades would be a perfect opportunity in which to take movies. I would try to get the closest and best pictures that I could of President Kennedy in particular.

I was able to take movies of two of the parades in 1962. During the first parade, I only had a standard lens so I was unable to take the close-up of picture that I really desired. Then I purchased a telephoto lens for the movie camera which enabled me to get much closer pictures. The camera had a slow motion mode which worked by speeding up the number of frames taken per second. As soon as President Kennedy came into view, I would set the camera into its slow motion mode.

More recently, I transferred the movies from the 8mm film to a DVD. An edited version of my movies of President Kennedy can be viewed online at www.gerstman.com/Kennedy.htm. A still clip from one of my films is shown below.

President John F. Kennedy and King Hasson II of Morocco
in a motorcade, Washington, DC 1962. ©George H. Gerstman

I kept and used my Revere movie camera for many years. In fact, prior to my purchase of a video camera around 1978, I always used my Revere camera for taking movies. I value these movies and I have saved them by transferring them to DVDs. No longer is it necessary to set up a movie projector and screen and thread the 8mm film through the projector in order to see the movies.

Super 8

In 1965, Kodak introduced a new 8mm film format called "Super 8." From then on, most people purchasing an 8mm type movie camera would purchase a Super 8 camera. The manufacture of standard 8mm cameras dissolved.

The Super 8 format had several advantages over standard 8mm. For example, it came in a plastic cartridge containing 50 feet of the Super 8 film. You would simply insert the cartridge into a Super 8 camera and you could then run the entire 50 foot length. There was no need to turn the cartridge over after using 25 feet, as with an 8mm magazine. A 50 foot length of film would run for about three minutes.

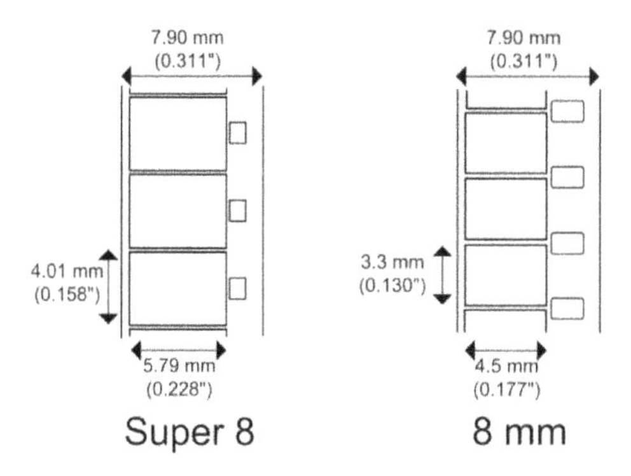

Comparison of Super 8 film with Standard 8 film
Picture credit: Max Smith

Also, the sprocket holes were smaller and spaced further apart, providing a substantially larger picture frame. Additionally, the cartridge was notched so that when it was inserted into the movie camera, the movie camera could determine the speed rating of the particular film.

Sound Movies

Super 8 cameras became the cameras of choice prior to about 1980, when video cameras started to become popular. Although most

Super 8 cameras did not provide sound, Super 8 sound film was sold in the 1970s. The film used a magnetic stripe on the perforation side to form a soundtrack. The sound film was used with special Super 8 sound film cameras.

There were 16mm and 8mm sound cameras which used 16mm and 8mm film having a soundtrack, but these were rare and used mainly by professionals. In 1973 sound cameras, such as the Kodak Ektasound Super 8 movie camera, became available for amateur use.

Kodak Ektasound 140 Super 8 sound movie camera

Although regular (silent) Super 8 cartridges could be used if no sound was required, the Super 8 sound cartridges were larger in order to accommodate the camera's recording head that engaged the magnetic soundtrack stripe on the film.

In 1968, prior to Super 8 sound film and Super 8 sound cameras, Phil Baron, an acquaintance of mine, invented a unique sound system for movie cameras and projectors without having to use a soundtrack on the film. It could be used with any type of motion picture camera: 16mm, 8mm, Super 8 or even 35mm professional studio cameras. There was no need for the camera to be battery operated. It could be a spring wound camera.

The sound would be recorded on one track of a stereo tape recorder from a microphone. The other track would record pulses that were generated by a small inductor device positioned adjacent one of the moving gears within the camera. Those pulses would aid in synchronizing the recorded sound to the movement of the film while the film was being projected. The movie projector used would not have to be a sound movie projector. The sound could be played, using the recorded tape, through any sound system.

I procured US Patent No. 3,492,068 for the invention. Phil and I formed a joint venture and we offered the invention for sale to camera companies. In 1970, shortly after the patent issued, the invention was sold to Bell & Howell.

Movie Lighting

While we have now become used to taking digital video indoors using available light, most people who have not taken movies using film probably do not realize that indoor movies required a significant amount of lighting. Using only available indoor light would generally result in a very dark movie. Kodachrome movie film, that was generally used, was relatively slow and was designed for outdoor light conditions.

When I took indoor movies during the 1960s and 1970s, I used two types of movie lights. Both were required to be connected to an electrical power receptacle. The first was a light bar which held four GE Photoflood lamps. The light bar was somewhat portable. It could be folded and placed it in a case for transport.

The light bar was much heavier and larger than the camera and was very cumbersome to use, although it provided a relatively even light for the picture. Typically, a second person other than the photographer would be in charge of holding the light bar during filming. If movement of the light bar was unnecessary it could be connected to a tripod.

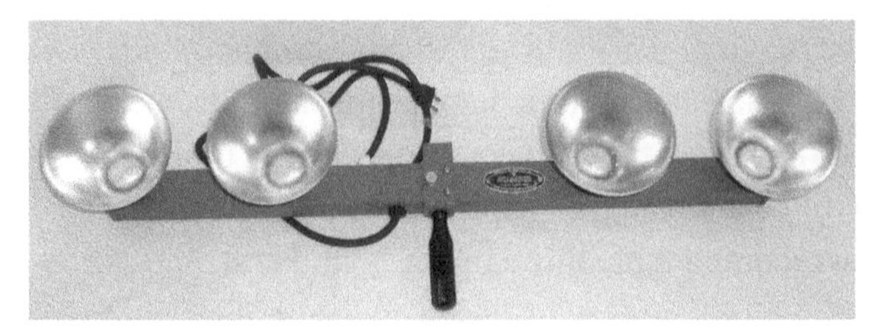

Crownlite Deluxe Foldmaster movie light lamp unit with four GE Photolamps

The other type of movie light was a single, powerful sealed beam unit, using a high intensity lamp. I had a Sylvania Sun Gun which used a 625 watt quartz lamp. It was relatively light and could be carried by its handle or connected via bracket to the movie camera. It had to be plugged into an electrical power receptacle which hindered portability. I recall that the subject couldn't get too close to the movie light because after a couple of minutes it would become dangerously hot!

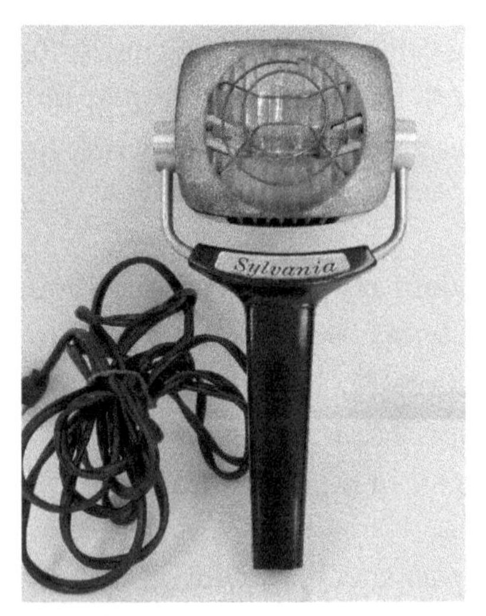

Sylvania Sun Gun movie light

CHAPTER 10

Taking Videos

One of the things that I liked best about taking movies was that my movie camera was very compact and all I really had to do when taking movies outside was point-and-shoot. Although I felt that video had many advantages over movies, at the beginning it was somewhat troublesome to take video.

EARLY VIDEO CAMERAS

Before I discuss why it was troublesome to take video, I want to mention the significant advantages of video. Early video was recorded on tape. In contrast with motion picture film, the video tape needed no processing. You simply recorded your video on video tape with a video camera and the tape could easily and immediately be played using a video cassette recorder (VCR) attached to a television set or any kind of television monitor. In this manner, you could watch your video momentarily after it was taken. There was no need to set up a movie projector or to set up a projection screen. All you had to do was insert the video cassette into the VCR. VCRs are discussed in an earlier chapter in this book.

Instead of a movie film which ran for only about three minutes in the camera, a full-sized video tape cassette would enable two hours of recording or even longer than two hours if lesser quality was suitable. The video recording would be in color with sound perfectly synchronized to the video. If there was a section of the video tape recording that you did not like, you could overwrite it. Or if you were satisfied with the tape you could remove a tab to prevent the tape from being overwritten.

Instead of using chemical imaging as with camera film and movie film, video tape used magnetic recording derived from an image sensor contained within the video camera. It really operated similarly to conventional tape recorders except that the video camera provided a video signal in addition to an audio signal on the tape.

Taking video sounds great! How could it be troublesome? The trouble I am referring to was the difficulty of carrying the video equipment that was necessary for taking video at the beginning. You had to carry a heavy VCR and a large video camera. To understand this, you have to recognize that the VHS type video cassette used in the 1970s through the 1990s was relatively large. It was approximately 7 inches long, 4 inches wide, and one inch in height.

A VHS videocassette

I purchased my first video camera around 1979. At that time, and until the early-1980s, video cameras did not have a provision for receiving a video cassette. Instead, the videocassette had to be inserted into a heavy VCR and a large video camera had to be connected via cable to the VCR. The VCR, not the camera, handled the recording of the video.

An RCA Newvicon video camera, 1982

The video manufacturers recognized that in order to take video, you needed both a video camera and a separate video recorder. They introduced portable video recorders to which the camera cable would be connected. The portable video recorder contained a rechargeable battery.

You could usually purchase a portable video recorder combined with a separate tuner which gave you the ability to record broadcasts using the tuner. The portable video recorder would be separated from the tuner during the taking of video and would be carried by the video photographer with the camera. However, these portable video recorders were relatively heavy, weighing over 10 pounds.

RCA portable VCR and tuner. The detachable portable recorder is on the right.

I had an RCA portable VCR like the set shown above. I don't recall ever using the tuner, so I kept the portable recorder portion separated from the tuner. I had a leather case for the portable recorder portion. I kept the recorder portion in the leather case because the case provided full access to all the inputs, controls, and the video cassette receiving slot of the recorder. The leather case had a shoulder strap which enabled me to carry the recorder over my shoulder. Although I considered it a hassle to carry all of the equipment during a video shoot, it was well worth it!

Woman using video camera
with portable VCR

Even if no television monitor was available, I could play back all or a portion of what I recorded using the playback controls on the portable VCR and viewing the playback using the viewer of the video camera. The video camera viewer also showed what I was taping during the taking of the video, and it performed as a very small monitor.

Taking videos aided me significantly in my work as a patent attorney. On many occasions, when I met with clients in their offices and factories, I was able to record technical material that would have been difficult to describe in writing. For example, if I was learning about the client's new invention, I would interview the inventor. Using my video camera equipment, I would take video of the inventor showing and describing his or her invention. This enabled me to record numerous details which would be extremely difficult to write as notes in a yellow pad. In addition, I found that most inventors thoroughly enjoyed having themselves describe their inventions on video tape.

When I traveled with my video camera and portable VCR, I carried it within a very sturdy case. The total weight was substantial and it was somewhat cumbersome to use, but I was enamored by the ability to take video at will. I look back at when my brother and I were teenagers and we saw ourselves on a black-and-white television monitor at the NBC studio where Dave Garroway hosted *The Today Show.*

In 1980, when my video equipment was fairly new, I decided to bring it with me on a vacation to Maui, Hawaii. Notwithstanding its weight and bulk, I carried the camera and portable VCR with me through the airlines and throughout the many days of activities. I took video of almost everything! While prior to video I would have 8 mm silent movies of relatively short segments of my activities and the scenery, I now had hours of my activities and scenery on video tape. I realize, however, that I greatly overdid the amount of time per scene and the tapes needed to be edited substantially. I transferred the video tapes to DVDs and I can easily run past the portions that appear boring! My children tell me that my recording of the Hawaiian sunset over the beach could be sold as a replacement for sleeping pills.

The Camcorder

The companies that manufactured equipment for taking and recording video recognized the difficulty in hauling around a video camera connected via cable with a heavy portable video cassette recorder. A much more compact system was needed. They developed what they called a "camcorder," which was a combination video camera and built-in video cassette recorder.

In order to provide a reasonably sized camcorder, the manufacturers introduced a new video cassette in 1983, designated VHS-C. The "C" stood for "compact." This video cassette was much smaller than the conventional VHS cassette. It could be played on a conventional VHS video cassette recorder using an adapter.

A VHS-C videocassette

The downsizing of the videocassette opened a new and exciting market in which consumers could purchase a camcorder that was easily portable and did not require a separate power pack. By using such a small cassette, the camcorder could incorporate an internal video recorder with playback, a rechargeable battery, a microphone, and a complete camera and imaging system.

A JVC Super VHS-C camcorder Model GR-S70U

I obtained a JVC camcorder, similar to the one shown above. It was lightweight, extremely easy to use, and for that time took amazingly good video. There was no high definition at that time, but the video was comparable to broadcast video. The camcorder was small enough so that I could carry it in a litigation case with other of my papers when I visited a client's offices or factory.

The viewfinder was a video monitor which showed the exact video that was being taken, although it was in black-and-white. Using the playback controls, I could review the video or portions of the video using the viewfinder as a monitor and I could overwrite any portion of the video that I thought should be deleted.

Although I had a VHS adapter for the VHS-C cartridge, enabling me to use the small cartridge in any conventional VHS VCR, I rarely used it. I found it very easy to use the camcorder itself to play back the recorded VHS-C tapes on a television set or television monitor. I would simply attach the appropriate cable to the output of the camcorder and into the input of the television, and then use the playback controls which were on the rear of the camcorder.

During the 1980s, all of the recording was analog and camcorders were becoming more and more popular. When you would go to a graduation ceremony, it seemed like at least half the parents would have a camcorder for recording the ceremony. Motion picture cameras were becoming obsolete.

In the mid-1990s, a new format was introduced to the public, enabling camcorders to shrink much further in size and weight. This new format was called Mini DV. Recording was digital and the Mini DV video cassette was considerably smaller than VHS-C cassette. It was approximately 2 ½ ×2× ¼ inch.

The smallest Mini DV camcorder that I saw was in an audio-video store in Highland Park, Illinois. It was so compact and light, and had every feature that I could imagine. It was a JVC camcorder similar to the one shown below, and I couldn't resist purchasing it. It was small enough to fit in a coat pocket and to me it was a precious jewel!

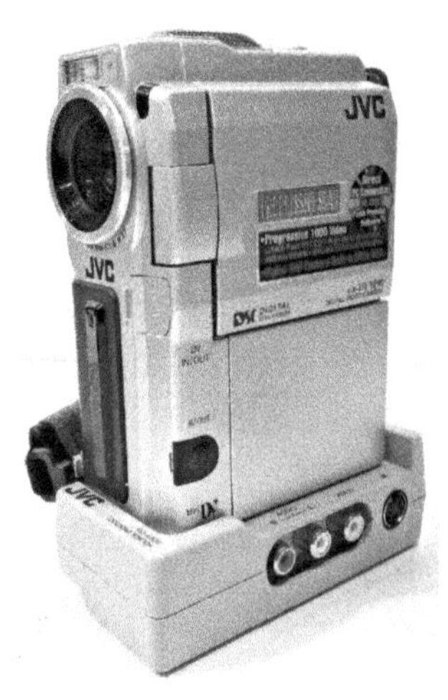

JVC Cybercam GR-DVM90 Mini DV camcorder
in its docking station

The JVC camcorder weighed only about 1 pound. It had a 2 ½ inch swing-out color LCD display for reviewing the video, an electronic color viewfinder, a 10X optical zoom, and a 200X digital zoom, among many other wonderful features. As with earlier camcorders, you could play the recorded tape directly from the docking station to a TV monitor. Or, if you didn't have a TV monitor available you could view the recorded tape on the swing-out LCD display without using the docking station. Although the resolution of the video recording was not in high definition as we now define it, the resolution was excellent.

The Flip

I thought camcorders had become as small as possible. But I was wrong. In 2006, the Flip camcorder was introduced. It was designed for people who just wanted to point and shoot video, without the complexities of other modern camcorders. It was so simple and small enough that you could carry it in your jeans.

Flip camcorder PSV-552 with its
USB plug in its flipped-out position

The Flip camcorders used a flip-out USB connector enabling you to connect the camera directly to a computer. The controls were extremely simple, enabling starting and stopping video recording, viewing videos and zooming. Some of the models recorded in high definition. The later models had an HDMI connection for use with TV displays.

At the same time, digital cameras with the ability to record video on memory cards as well as still pictures were becoming popular on the market. My Sony digital camera discussed in an earlier chapter of this book is an example. These digital cameras had a much larger image sensor and better lens than the Flip camcorder. They were able to provide an extremely high-quality video as a result of the large image sensor and high-quality lens. These digital cameras often provided better video than most popular camcorders as a result of the larger image sensor used by the digital camera. Many of the camera models had interchangeable lenses which enhanced their versatility.

While these cameras are relatively bulky, mirrorless digital cameras and DSLRs are still widely used by amateurs and professionals for videography. Modern digital cameras ordinarily enable high definition resolution, up to 4K.

The Iphone With Video

In my opinion, the shrinking of the video camcorder took its most exciting turn when Apple introduced a video camera into the iPhone 3G in 2009. This feat resulted in a camcorder that weighed less than 5 ounces, had a 3MP camera, and a 3 ½ inch display. Two years later, high definition (1080p) video recording was introduced with the iPhone 4s. In 2013 Apple introduced slow motion video on the iPhone 5s and optical image stabilization was introduced on the iPhone 6 the following year.

Each year the display screen increased in size and the number of megapixels of the camera increased. In November 2017 the iPhone X was introduced with a 5.8 inch OLED display screen, two 12MP cameras and 4K video recording. The 4K resolution is astounding and various resolutions are selectable in the Settings menu shown below.

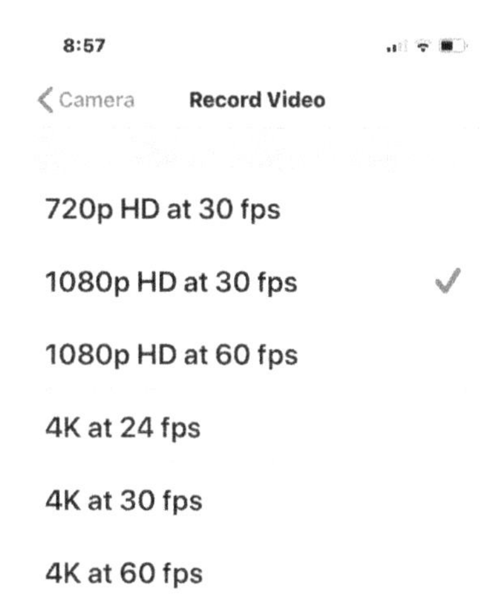

A partial screenshot of the iPhone X camera's resolution settings

A 29 second hand-held video that I took in NYC with 4K resolution at 60fps, using my iPhone X, can be viewed online at www.gerstman.com/4Kvideo.htm.

The iPhone 11, which was introduced in September 2019, has additional video recording features. It enables you to make a time lapse video. It has continuous autofocus and playback zoom. You can take still photographs while you are recording in 4K video. If you are about to take or are taking a still photograph but decide that a video of the scene would be better, you just have to hold the shutter button down and a video recording will proceed.

One of the things I like best about video recording with a smartphone is how easy it is to send the video around the world. With practically no effort, I can email the video, text message the video, upload the video to YouTube, Vimeo, Skype, and so many others, as well as being able to stream the video while it is being taken. And all of this can be done without requiring any external equipment -- just a handheld smartphone connected to the Internet.

CHAPTER 11

The Internet

I never heard of the Internet until I was at least 50 years old. Prior to around 1990, it was a relatively obscure network of computers used by academia.

BEFORE THE INTERNET

What did we do prior to the 1990s to obtain information? There were no search engines so we primarily relied upon encyclopedias and libraries. Most of the families that I knew each had a set of encyclopedias. We had the World Book Encyclopedia and others had the Encyclopedia Britannica, or both. At the time, there were actually door to door salesmen who would sell encyclopedia sets to families. Because much of the content was constantly changing, the encyclopedias constantly needed to be updated in order for the information to be relatively current. The encyclopedia companies sold annual supplements and of course they tried to convince people to buy completely new sets of encyclopedias periodically.

In general, the encyclopedias provided a relatively limited amount of information about a limited variety of subjects. For more detailed information, it was necessary to do the research at a library. Typically, you would go to a local library, check for the subject in the card catalog, and then try to find the books relating to that subject. Often one or more of the books were missing. If you found what you were looking for in one of the books, and needed a copy of the information, you would pay the librarian for photocopies.

If the subject matter that you were searching was more complex, such as certain arcane scientific subject matter, it would be advantageous to go to a scientific library if there was one in your city. Researching a subject could be a tedious endeavor and handling such research was an art itself.

Shopping for anything was very different prior to the Internet. Amazon didn't exist. eBay didn't exist. Webpages didn't exist. Buying something required going to the place where it was being sold or else referring to advertisements in the media or classified advertisements in a newspaper or the yellow pages of the telephone directory. Prior to 1970, credit card information was not accepted over the telephone. Even if you wanted to purchase something from a media advertisement, you would have to mail a check or money order to the merchant.

Finding unusual items was really a difficult task. For example, I collected scientific antiques. One category of my collection was vintage radios and televisions. In order to find these, I would have to go to special shows, flea markets and antique stores. Finding a particular vintage radio could be extremely difficult and time-consuming. If it wasn't at a nearby location I might never find it.

Likewise, finding a particular brand of a staple article could be extremely difficult if it was not being sold locally. You really had to depend upon the article being carried by a local store and if you lived in a rural area, shopping was a tremendous chore.

Think how simple it is now to find the objects of your desire! If you're seeking a certain vintage radio you can Google it or enter the name in the eBay search box. Looking for a particular brand of a staple article? Just Google it.

The World Wide Web

It wasn't the Internet per se that changed the world. It really was the World Wide Web and the use of the web browser on the Internet. That is what enabled us to have interactive webpages, providing global visual communication.

In 1991, my law partner Terry McMillin and I decided that it might be useful to connect our computers to the Internet. We knew very little about the Internet and we thought it would be a good learning experience. After doing some research, we found that we needed an Internet service provider (ISP), and a telephone line for a dial-up connection with the Internet service provider. We also needed a modem connected to the telephone line and to our computer. The modem would decode the audio signals from the telephone line and provide the decoded information to the computer. In this manner, we could have a telephone connection to the Internet service provider, which effectively connected us to the Internet and enabled data communication.

We found an Internet service provider named Delphi, and we began a subscription to Delphi. This was before there were websites or webpages and our Internet experience was primarily with chat groups where we would obtain and convey information. There were groups for every category of information. If you wanted information about organic chemistry, there were a number of organic chemistry groups. If you wanted information about photography, there were numerous photography groups. During an Internet "chat" you would just see text on the display screen, somewhat similar to text messages.

Although it must seem boring to view only text while on the Internet, it actually was quite exciting. We were able to communicate with people around the world who could provide interesting information. In my legal practice as a patent litigator, I often needed technical information that was not readily available from my client.

On many occasions I was able to post a question and obtain the information over the Internet from a person skilled in the particular field. Usually I would receive a number of responses.

I remember that in the early 1990s we used a program on the Internet called Gopher. Gopher became our favorite source of information. I recall doing research in connection with a patent that used the term "stagnation pressure." The meaning of the term became very important to my research but I was unable to find its definition anywhere. I posted a query on Gopher, asking if anyone knew the definition. Within a couple of hours, I obtained several useful replies from scientists around the world. The most detailed response was from a PhD scientist at NASA.

We have become used to high-speed broadband data communication and Wi-Fi. In the early to mid 1990s these did not exist. We used a telephone dial-up that was many magnitudes slower than the present download and upload rate. We first had to dial the telephone number of the Internet service provider. Very often the line was busy and we were unable to reach the Internet service provider for many minutes. Once we had a telephone connection, we heard noises similar to fax connection noises, and waited for the handshake. Once the handshake occurred, we were connected to the Internet.

Although I now use broadband and my download speed is approximately 400 million bits per second, when we used dial-up telephone service in the 1990s the download speed was a maximum of 56 thousand bits per second, almost 10,000 times slower!

In late 1994 I learned that a web browser was being tested that enabled people on the Internet to see webpages that contained text, pictures and sound. It was designed by a company named Netscape. On the media there were discussions of various companies that built websites enabling people on the Internet to access information concerning these companies. I discovered that a new start up, called

America Online or AOL, was an Internet service provider that enabled access to the World Wide Web. The World Wide Web was and still is the medium on which websites and webpages exist.

I subscribed to AOL and was fascinated to use the AOL-modified Netscape browser to view the various websites on my computer. AOL had a feature by which I could build my own website and broadcast it on the Internet. This greatly intrigued me, and using an AOL template in early 1995 I constructed a simple website for my law firm. However, I felt that the website looked amateurish and to access it required going to an AOL webpage. Using AOL became a pain because their telephone lines were constantly busy. It was time to subscribe to a new Internet service provider. Also, I decided to learn how to build a nicer website and use my own domain name.

In mid 1995 I contacted a local Internet service provider named Interaccess and became a subscriber. I purchased some books about building websites and webpages, using HTML code. I also requested Interaccess to obtain for me the domain name "gerstman.com" hoping it was available. Now you are able to find out immediately if your proposed domain name is available but in 1995 it took several days. It was available and it was secured for me through a company which still exists, called Network Solutions.

Interaccess became my web host, meaning that my webpages would be uploaded to Interaccess servers. Then when anyone viewed any of my webpages, the pages would be downloaded from the Interaccess servers.

I purchased a program named WebEdit that aided me in building webpages for my new website. Using this program, I was able to enter HTML code and then see what a webpage would look like based upon the code. Once the webpage looked satisfactory, I would upload it to the Interaccess server. Although I have moved my gerstman.com account to a new web host (Yahoo) some of my original webpages from the 90s are on Yahoo's servers now.

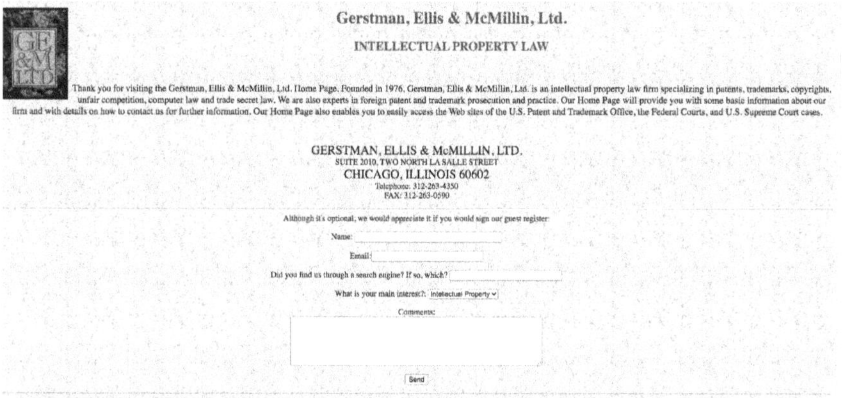

Screenshot of the early home page of my law firm's website, which I built in 1995.

Downloading pages from the Internet required a substantial amount of patience. Because of the slow Internet connection using a dial-up telephone line, in the late 1990s it wasn't unusual to take as much as 30 seconds to download a webpage. In building a webpage in the late 1990s, it was important to keep the page from being "heavy." A "heavy" page would contain so many bytes of data that it could take a relatively long period of time for people to download the page. This is no longer a problem because of the use of fast broadband Internet connections. But when you were using a telephone dial-up Internet connection at less than 56 kilobytes per second, the pages had to be kept "light" to decrease the download time. Also, storage was at a premium and the web host would add a surcharge for heavy pages requiring more storage.

One way to keep the webpage light was to use no photographic images or to use low resolution images. Almost everything other than text required a relatively large number of bytes. In 1995, so far as I was aware, video, which is heavy with bytes. was not used on webpages. However, being curious I decided to experiment with video and in 1995 I was able to create videos that I used on webpages that I built.

To place a video on a webpage I took a video with my video camera. I then converted my video into gif format creating a number of consecutive pictures. I then introduced a small gif viewer into the webpage, using an appropriate HTML code. At a telephone/modem access rate of 56 kilobytes per second, the pictures in gif would be shown at about 10 pictures per second. The result was a viewer on the webpage showing a video comprising the gif photos moving at a rate of about 10 per second. Later in the 1990s a simple HTML tag was introduced for inserting video in a webpage.

In order to gain more education concerning webpage construction, I would view the HTML source code that others were using. Using the Netscape browser or the Internet Explorer browser, when a webpage was on the screen you simply would click on View, then Source, and the source code was displayed. On the Chrome browser you click on View, then Developer, then Source Code, and the source code is displayed.

Although the Netscape browser, called Netscape Navigator, was first and was extraordinarily successful, it is now obsolete. Microsoft bundled its own browser, Internet Explorer, with its Windows software and Microsoft's Internet Explorer became dominant. Now there are many good browsers available, including Safari, Firefox and Chrome. My preference is Chrome, a Google product.

In the early days, the source codes were so simple. Many pages contained less than 100 words of source code. In those days, we typed in the HTML code without using machine code automation and pre-existing templates. The web editing programs were helpful, but you still had to know how to use the HTML tag and code structure in order to build a webpage. I then learned JavaScript programming and was able to use it in a certain of my webpages.

Recently I built a new website using a Wix.com template. The Wix website builder using online drag-and-drop tools made it very simple to build an attractive and effective website. It's very easy to modify the website and to add new webpages. Everything you create can be viewed immediately on the web. Interestingly, if you view the source code it looks extraordinarily complicated but by using a template with drag-and-drop tools you do not get involved with any of the coding.

Domain Names

Earlier in this chapter I mentioned the obtaining of the domain name "gerstman.com." There are many tales that can be told about domain names.

The value of a particular domain name can be immense. Think of amazon.com, google.com, ford.com, united.com and thousands of other domain names that are the names of significant companies throughout the world. Don't think that all of these companies simply registered the desired domain name. Many of the desired domain names were already registered by others and had to be obtained through legal action and/or monetary payments. During my legal career, I was involved in obtaining and protecting domain names for many of my clients and I am pleased to discuss some of this activity.

In 1994, when the World Wide Web became very active and a browser was available for viewing websites, many persons and companies recognized the value of registering one or more domain names. At the time, most domain names were available and easy to register. Network Solutions was the accredited registrar back then. At first there was no charge for registering a domain name but in 1995 Network Solutions begin charging yearly registration fees.

Certain people saw an opportunity to register the names of famous companies as domain names, while those names were still available and before the respective companies recognized the need to obtain a domain name. For example, in 1994 an individual living in Champaign, Illinois named Dennis Toeppen registered over 200 famous names as his domain names. These names included ussteel.com, unionpacific.com, neiman-marcus.com, britishairways.com, americanstandard.com, and crateandbarrel.com.

In 1997, Gordon Segal, the president of my client Crate & Barrel, decided that Crate & Barrel should have a website. I found that the domain name crateandbarrel.com was not available but was registered to Toeppen. I contacted Toeppen and was not surprised to find that he demanded ransom for the domain name. People who had registered domain names with the names of famous companies and sought ransom were called "cybersquatters." There had recently been cases in which cybersquatters (one of them being Dennis Toeppen) were enjoined by courts from maintaining the domain name and were ordered to transfer the domain name to the proper tradename owner. I filed a lawsuit in federal court against Toeppen for cybersquatting. It didn't take long before the case was settled and the registration was assigned to my client.

Another trick used by shady companies is to take a slight variance of a famous name and use it for capturing business. In this manner, if you happen to type in the incorrect spelling of a company's name inadvertently, you will be misdirected to a website controlled by the shady company. To alleviate this, I would advise my clients to register many variations of their names. If you accidentally type in one of these variations, you will be directed to the correct home page. For example, type in "amzon.com" and you will be directed to "amazon.com."

Over the years, primarily between 1996 and 2006, I filed numerous federal lawsuits against companies that used my clients' domain names improperly. In 1999 Congress passed the Anti-cybersquatting Consumer Protection Act. This made it even easier to stop the pirates.

My client Blublocker Corp., the famous sunglasses company, had the domain name blublocker.com. In 2005 a competing Texas sunglass company started using the domain name blueblockers.com ("blue" instead of "blu"). This was a classic domain name misuse. I filed a lawsuit in Chicago and the registration was transferred to my client promptly. In every lawsuit that I filed for improper use of the domain name, the defendant either defaulted and I obtained a default judgment or else the defendant settled promptly by transferring the domain name registration to my client.

The value of a domain name to a company is hard to calculate. It may actually be worth millions of dollars. A client of mine, a large international public company, was unable to obtain the domain name that it wanted because the domain name was previously registered to a small company in a different type of business that had the same name. I learned that the small company was purchased by a Canadian conglomerate and that the domain name was no longer being used. I advised my client of this and I was authorized by my client to try to purchase the domain name from the Canadian conglomerate for up to $500,000. I contacted the president of the Canadian conglomerate and told him that my client was willing to pay $24,000 for the domain name which his company was no longer using. He said that his company would sell it to my client for $25,000. I told him that I would contact my client and get back to him! After a few minutes, I called him back to accept the deal, and the transfer was completed within a day.

I find it interesting to see some of the old websites used by various companies and organizations. Using domain names, there is a search engine that shows historical websites. I found this to be very helpful in trademark litigation. The search engine is archive.org, and when you enter a domain name you have the ability to see examples of the associated websites as far back as 1996. Often the links in the early websites continue to link to the associated early webpages.

CHAPTER 12

The Calculator

The calculator is a machine that has evolved in an unbelievable way over the years. Presently, when we want to add, subtract, multiply or divide, and use other math functions, we can go to the appropriate app on our smartphone. It's so simple and the accuracy is phenomenal. This is very different from the past.

When I was in elementary school, high school and college, there were no desktop or handheld electronic calculators. In the 1950s, while I was in college, there were analog and digital computers but these were available only to large institutions and were extremely massive and expensive.

The calculators used in the 1940s, 1950s and 1960s were really adding machines. They could only add and subtract in a single operation. You would enter the numbers on the keyboard to be added and pull a crank. If you wanted the numbers to be subtracted, the numbers would be entered and the driveshaft of the machine would be reversed when you pulled the crank. To multiply numbers or divide numbers, you would need to repeatedly add or subtract numbers.

Monroe High Speed Adding Machine, circa 1940s

The Monroe adding machine, shown above, is typical of the calculators used during the 1940s 1950s and 1960s. They were approximately the same size as typewriters at that time. Some had paper tape outputs.

During the time I was in elementary school and high school, because small electronic calculators were unavailable, we were required to compute arithmetic and mathematical functions by hand. That included addition, subtraction, multiplication, logarithmic functions, exponential functions, trigonometric functions, and more. For many of the mathematical functions such as logarithmic functions, we would need a sheet or book containing a printed logarithmic table.

In 1956 I learned how to use a slide rule. At the time, a slide rule was a necessity to the mathematics used by mathematicians and engineers. In that year I began my study of electrical engineering at the University of Illinois in Champaign-Urbana. The first thing that every engineering student purchased was a slide rule. I bought a Post Versalog slide rule and used it constantly throughout engineering school.

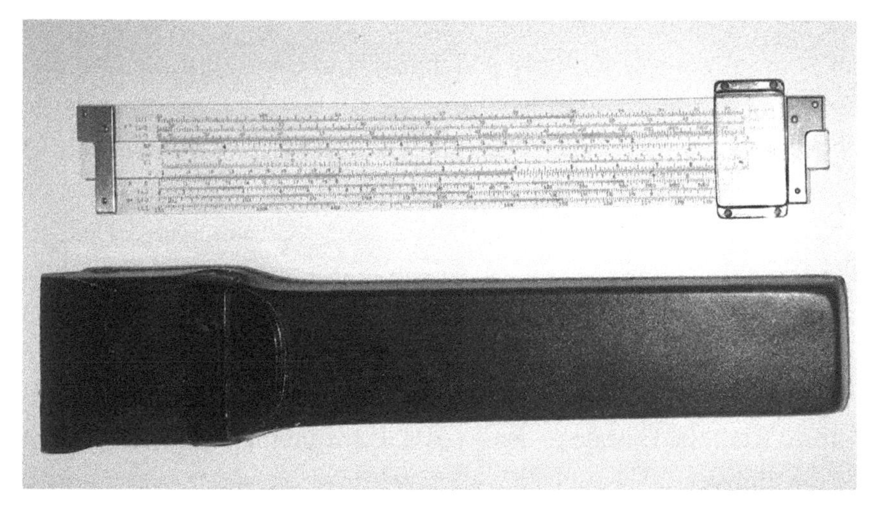

Post Versalog 1460 bamboo slide rule and case

It was easy to detect the engineering students on campus. They were the ones who had their slide rule cases attached to their belts like a holster, enabling them to whip out the slide rule whenever it was needed. One engineering student was really gung-ho. He carried the slide rule in a case attached to a belt that crossed his chest, like a bandolier worn by the Mexican revolutionary Pancho Villa!

A slide rule is a powerful tool having only a few parts. You cannot add or subtract with it but you can do a myriad of other math functions.

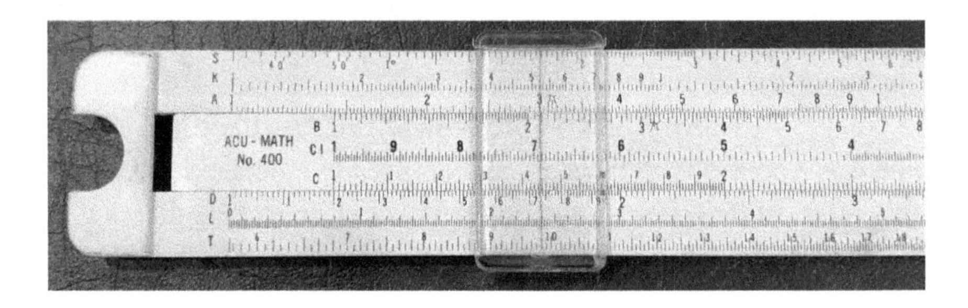

Examples of some of the many math functions that can be performed on a slide rule are shown in the photograph above. Note the lettered scales S, K, A, B, etc. on the left side. Although there are several ways to multiply and divide, I prefer to use the C and D scales. If I want to multiply 12 times 2, I place the 1 on the C scale above 12 on the D scale. I then see that 2 is aligned with 24 on the D scale, which is the product of the multiplication. If I had wanted to divide 24 by 2, I would place the 2 on the C scale over 24 on the D scale and the quotient (12) is shown under the 1 on the C scale. If I want the square root of 3, I align the hairline on the slider with 3 on the A scale. Dropping down to the D scale, you can see that the hairline is aligned slightly above 1.73. To get the square of a number, you go from the D scale to the A scale. If I want the base 10 logarithm of the number 3, I would slide the hairline of the slider over 3 on the D scale and scale L would show the base 10 logarithm of 3 as 0.477 (it's helpful to use a magnifying glass!).

So far as I was aware, until the late 1960s the only electronic calculators available were large computers which are discussed in an earlier chapter of this book. If you wanted a calculator on your desk, you were limited to a mechanical adding machine and/or a slide rule. However, as transistors and solid-state circuits became more prevalent in the 1960s, there was a movement to use them in the manufacture of an electronic calculator that would be suitable on the desk.

In the late 1960s, I became aware of electronic calculators that were approximately the size of typewriters. These calculators not only could add and subtract, but they also had the ability to multiply and divide. The display was often a series of Nixie tubes and many of the calculators also had a paper tape output. They had to be plugged into a source of AC power and were not battery-operated. They were priced at about $1,000 at the time.

Burroughs Model C 3660 16-digit Nixie tube calculator

One well-known manufacturer of electronic desk calculators was Burroughs, which for many years had manufactured business equipment from mechanical adding machines to mainframe digital computers. Burroughs invented Nixie tubes which resemble vacuum tubes containing numbers that glow a reddish-orange color when energized.

One of the most exciting new products was introduced in the fall of 1971. It was the "pocket" calculator, a battery-operated electronic calculator that could fit in the palm of your hand (or in your pocket) and could add, subtract, multiply and divide. The earliest US-made pocket calculator was the Bowmar model 901B which used a keyboard manufactured by Texas Instruments and a display comprising a series of red LEDs. At the time, the Bowmar pocket calculator cost about $200.

Bowmar 901B calculator
with attached charging cord, 1971

The company that first introduced the Bowmar pocket calculator to the public in 1971 was my client JS&A Group. In 1971, JS&A was a tiny company and its president, Joe Sugarman, was literally working out of his house. How was he able to scoop the earliest sales of this significant new device? He saw the calculator at a trade show before it was being sold to the public and he made them an offer. He told them that if they would give him an exclusive for a limited period of time, he would have the calculator for sale in advertisements in major publications within the next few days. No one else had the ability to compose an advertisement and publish it so quickly.

Bowmar accepted Joe's offer. Being an expert copywriter, Joe and his secretary created a camera-ready mail-order advertisement overnight and sent it to *The Wall Street Journal*, *The New York Times*, and other newspapers for immediate publication. The advertisement was a huge success and the Bowmar pocket calculator and JS&A were off to a great start.

Competition in the sale of pocket calculators was intense. Everyone wanted one because it made it so easy to calculate arithmetic functions. The prices of the parts, including the integrated circuits, displays and keyboards were declining and the prices of pocket calculators decreased significantly each year. With sales taking off, there were improvements, such as the use of LCD displays instead of LED displays in order to save battery life.

One of my favorite improvements was in 1975 when a scientific pocket calculator was introduced. This calculator not only handled arithmetic functions, but it also handled many other mathematical functions such as logarithmic, exponential, roots, and trigonometric. And it sold for under $25! I'm referring to the Texas Instruments TI-30 scientific calculator.

Competitive scientific calculators followed. The TI-30 calculator and other scientific calculators cost less than a good slide rule. They were far more accurate than a slide rule. They had so many advantages over the slide rule that the slide rule became obsolete.

Texas Instruments TI-30 scientific calculator, 1975

Now scientific pocket calculators are often used by elementary school, high school and college students. In many cases, they can be used on exams, even the SAT. They are so popular that you can even buy them at drugstores and it's not unusual to find a brand-new scientific calculator selling for only around $5.

The pocket calculators today generally have an LCD display and are often powered by solar or a combination of battery and solar. In addition to scientific calculators, graphing calculators are popular. Also, you can find every type of calculator imaginable as an app on your smartphone. One of the apps that is built into the iPhone now is an arithmetic calculator. If you want a scientific calculator on your iPhone, there are dozens to choose from. As an example, Calculator HD looks like the photograph below on the iPhone display. Among other things, it has 3D touch buttons, a dual-line display that shows the inputs and results at the same time, and a time and date stamped history tape which saves each calculation.

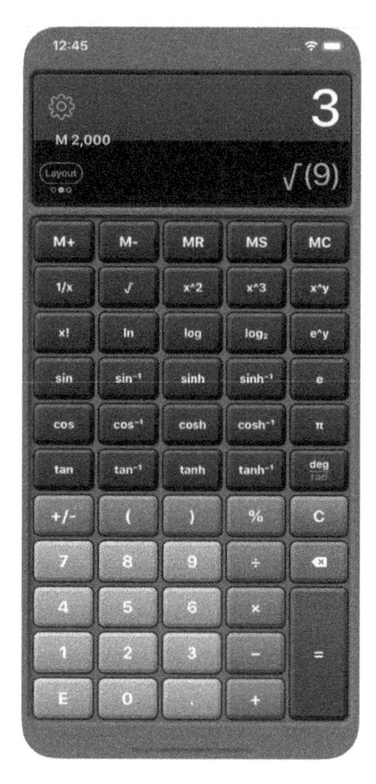

Screen shot of Calculator HD app
scientific calculator on iPhone

CHAPTER 13

Video Games

THE INCEPTION

Video games were born in the 1970s. Enthusiasm for video games and their growth has been phenomenal. It has been estimated that the global video game market will generate more than $175 billion in 2021.

Although I am not a video game aficionado at the present time, I was embroiled in the early adventures and progress of video games, from their beginning into the 1980s. In this chapter, I will discuss some of the early adventures in which I was personally involved.

Most people think that the first video game was PONG, sold by Atari. This is incorrect. The first video game was sold by Magnavox in September 1972, and it was called Magnavox Odyssey.

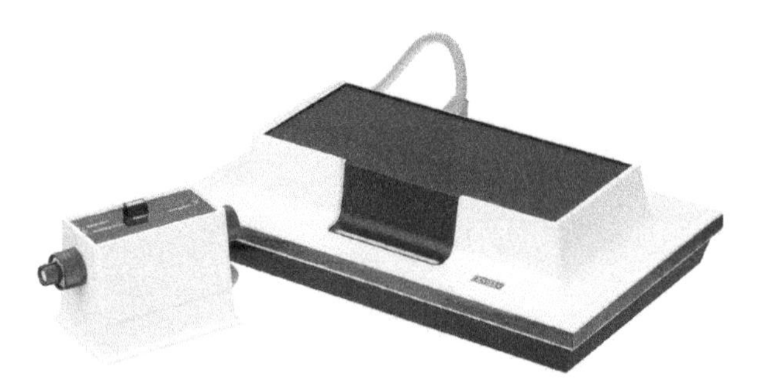

Original Magnavox Odyessy with one of the two game controllers, 1972

The Odyssey console was connected to the antenna terminals of a television set. It was the beginning of the video game revolution. It was the first device that enabled the consumer to interact on a television screen with the action that was being presented on the television screen.

The Odyssey sold for about $100 in 1972. The game in the box included the main console for connection to the TV antenna terminals, two game controllers, a switch for toggling the television set between its normal antenna and the Odyssey, a set of game cartridges, numerous overlays to stick to the television screen, and a number of board game supplies.

Original Magnavox Odyssey set, 1972

The picture that the Odyssey displayed on the screen was primitive. There were just three rectangular white dots and a white vertical line. There was no color or sound. Two players could each control a different one of the white dots and the system controlled the other dot. The vertical line could be moved horizontally to change the game from tennis to handball, for example. A 1972 promotional video showing the Odyssey being played can be viewed online at www.gerstman.com/odyssey.htm.

In early 1972, Nolan Bushnell, the founder of Atari, visited a Magnavox plant and had the opportunity to play the Odyssey tennis game demonstration. Apparently, this inspired Bushnell to conceive of a similar game in the form of a ping-pong game. He assigned the task of designing the ping-pong game to Al Acorn, an Atari engineer. A demonstration model of an arcade game was developed and proved to be very successful. Atari's PONG arcade game was introduced before the end of 1972 and it became the first arcade video game. At the time, the most popular arcade game was pinball but starting with PONG, the arcade video revolution began.

Original PONG arcade video game, 1972
Photo credit: Chris Rand / CC BY-SA/3.0
(https://creativecommons.org/licenses/by-sa/3.0)

PONG was similar to the tennis game on the Magnavox Odyssey, but slightly less primitive. It included sounds and the score was shown on the screen. Each of the two players had a joystick on the front of the console to control the vertical position of the player's paddle. Each player's paddle was on the opposite side of the screen and they hit the ball back-and-forth until one player reached the goal of 11 points and won. The paddles and the ball were simply rectangular white dots and the background of the game was black. A screen shot of the game in play (from a YouTube video) is shown below. The YouTube video showing the PONG gameplay can be viewed online at www.gerstman.com/pong.htm.

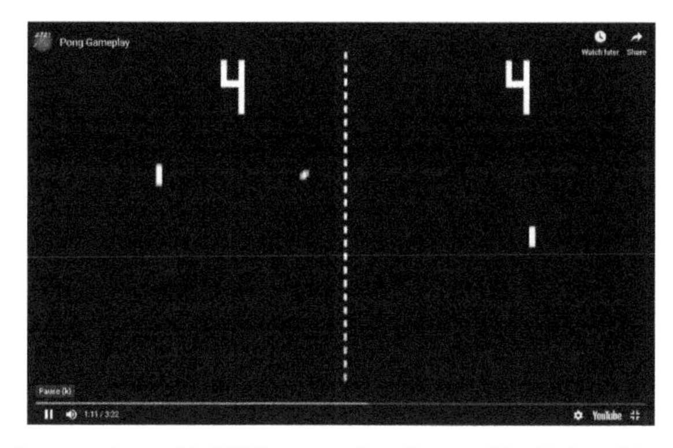

Screen shot of PONG gameplay from a YouTube video

When electronic game manufacturers saw how successful PONG was, they wanted to jump on the bandwagon. In 1973, a client of mine named Universal Research Laboratories (URL) decided to manufacture a video game similar to PONG that enabled four-person play instead of only two. URL entered into a venture with Allied Leisure Industries, a company that was already in the arcade game business.

URL wanted to protect its development somehow. At the time, the issue of protecting the intellectual property of a video game other than by a patent had not been considered. However, I thought it would be useful to claim a copyright on the printed circuit board as a "work of art." I told my client to place a copyright notice on the printed circuit board, which URL did in the name of Allied Leisure. URL also added a false circuit line to trap copiers.

In 1973, Allied Leisure introduced an arcade game named Tennis Tourney. It used the circuit board manufactured by URL and it was similar to PONG but it enabled four players to play the game. It was very successful but shortly thereafter Bally Manufacturing Corp., a giant in the arcade game field, introduced a less expensive four player tennis video game that was very similar to PONG and to

Tennis Tourney. URL obtained one of Bally's games and noted that the printed circuit board was identical to the Allied Leisure printed circuit board. It even had the false circuit line!

URL and Allied Leisure asked me to try to stop Bally from selling this game. Although I knew that this type of copyright issue had no precedent, I prepared a complaint and a motion for a temporary restraining order and filed it in federal court in Chicago. I asked for an immediate hearing and it was set for the next morning. At the hearing, the unique copyright issues were argued with Bally contending that a printed circuit board cannot be copyrightable. Although the judge did not issue the temporary restraining order, he set it for a preliminary injunction hearing within one week.

This was the first intellectual property case in the United States relating to video games. Unfortunately, the copyright issue was never resolved because prior to the preliminary injunction hearing Bally settled by taking a royalty-bearing license from Allied Leisure.

Magnavox had two patents relating to its video game technology. In 1974, Magnavox sued Atari, Allied Leisure, Bally and Chicago Dynamics for patent infringement. The case resulted in monetary settlements. My client URL, which was not sued, began manufacturing and selling a home-oriented video game named Video Action. It was in a consumer form like the Magnavox Odyssey, in that it was a relatively small console connectable to the antenna of a television set. Others, including APF Electronics, also began manufacturing video games for the consumer market. These home games were very popular and were sold by major retailers including Sears and Montgomery Ward.

In 1977 Magnavox sued my client URL, along with APF, Sears, Wards and other retailers for patent infringement. I was hired by those four and other defendants to defend them in the case which was in federal court in Chicago. As a result of my involvement in this lawsuit, I became very familiar with the technology and the people in the video game industry at a time when the technology was developing rapidly. After many years of pre-trial discovery and hearings, the cases were settled with licenses.

During the 1970s, the audiovisual quality of the video games improved greatly. The designing of a video game became an art in which software coders were the artists.

The popularity of arcade videogames exploded at the end of the 1970s and into the 1980s. One of the most popular video games in history was Space Invaders, introduced by Bally in 1978. Playing the game, you would gain points by shooting down, with a laser gun, waves of descending aliens. The Space Invaders gameplay can be viewed online at www.gerstman.com/spaceinvaders.htm. Space Invaders earned billions of dollars, giving a huge boost to the industry. Other iconic games from that time included Pac-Man and Donkey Kong.

In 1977, my client URL merged with an arcade game company named Stern Electronics, run by Gary Stern who was well known in the industry. Stern Electronics was a manufacturer of jukeboxes, pinball machines and arcade video games. I began handling the intellectual property matters of Stern Electronics.

A legal adventure started with Stern's first arcade video game named Astro Invader. Stern had an exclusive license in the United States from Union Denshi, a Japanese company. The software and circuitry would be developed in Japan and Union Denshi would send the circuit boards to Stern. Stern would build the arcade game with a cabinet, television monitor, circuit boards, speakers, player controls, and coinbox.

The game programs for all of the arcade games were stored on PROMs, which were programmable ROMs in the form of integrated circuits, carried by the circuit boards. These PROMs could be easily duplicated but at the time the Copyright Office was not granting copyright registrations for computer programs on PROMs or ROMs.

How We Stopped The Pirates

Stern wanted to protect its video games from copiers but there was no precedent for stopping copiers who sold video games that had the same look and gameplay. There was already an operator named Harold Kaufman, doing business as Bay Coin, who was installing bootleg arcade games called Zygon which used the same program and had the same display as Astro Invader. The games were being supplied by Omni Video Games, Inc.

Stern was able to obtain one of Omni's infringing video games. I took video of the infringing video game in action. Comparing the video of Stern's video game with the video of the infringing video game, they were identical.

I wanted to bring copyright infringement lawsuits in federal court against the infringers. In order to do this, I needed a copyright registration for the videogame. I decided to try to register the video game as an "audiovisual work," similar to the way a motion picture is registered. In other words, the copyright registration would cover the sights and sounds of the video game. A copy of the work to be registered must be sent to the Copyright Office. To that end, I took videos of Astro Invader in action. I requested expedited registration based on the infringement and sent the copyright application with two VHS cartridges containing the recorded video game in action.

The Copyright Office accepted my application and video tape recording and issued a copyright registration. I immediately filed a lawsuit for copyright infringement in federal court in Brooklyn against Kaufman and Omni and requested a preliminary injunction.

The defendants consented to the preliminary injunction pending the hearing, stating that they would argue at the hearing that a preliminary injunction would be improper.

Prior to the hearing, Stern introduced an arcade video game named SCRAMBLE, that was a blockbuster. Very shortly thereafter, Omni begin selling an identical videogame, also named SCRAMBLE. I immediately obtained a copyright registration for SCRAMBLE (as an audiovisual work) and I supplemented the complaint to add this infringement to the case. From then on, I would base my copyright infringement arguments on the SCRAMBLE game which was bringing in millions of dollars. The gameplay of the SCRAMBLE video game can be viewed online at www.gerstman.com/scramble.htm.

This was the first case in the United States in which a court was going to decide the copyrightability of the visual images displayed by a video game. Many representatives from the video game industry attended the hearing in Brooklyn. During the hearing, the testimony and cross-examination went very well. Among others, I called the president of the defendant Omni to the stand as an adverse witness. His testimony actually helped my case.

The preliminary injunction was granted with an excellent published opinion from presiding US Judge Eugene Nickerson. The industry was delighted. Omni filed an appeal to the United States Court of Appeals for the Second Circuit. This would be the first appellate case to consider whether copyright protection is available for the visual images displayed by a video game.

After Judge Nickerson's decision and before the hearing at the Court of Appeals I brought numerous other copyright infringement lawsuits in various areas of the country. At the same time that I filed each lawsuit, I would request a Temporary Restraining Order and an order requiring seizure of the infringing video games by the United

States marshals. By indicating in my request that the defendants may hide the bootleg video games and other evidence if they had advance notice of the lawsuits, I was able to obtain hearings before the judges without notice to the defendants. I took videos using my video camera and portable VCR for showing the video games in action to the federal judges. The Temporary Restraining Orders and seizure orders were always granted.

The first of those lawsuits that I brought requesting a Temporary Restraining Order was against an arcade in Coney Island. They had a bootleg SCRAMBLE video game. Stern wanted the industry to know that pirates would be sued and the bootleg video games would be impounded. I obtained the requested Temporary Restraining Order and the next day I hired a truck with two movers and was accompanied by a deputy US marshal to the Coney Island arcade. The marshal showed the seizure order to the arcade manager and the bootleg SCRAMBLE game was impounded, placed on the truck, and moved to a warehouse. I personally took photographs and prepared a news release. A story with photograph appeared in *Playmeter* magazine, which was the primary magazine for the coin-operated arcade game industry.

Subsequently the lawsuit was settled by the parties. The impounded game from Coney Island became Stern's property and Stern gave it to me. I had this SCRAMBLE game in my basement for many years and I became very proficient at playing it.

Then we learned that a number of bootleg SCRAMBLE games were placed in the Tulsa, Oklahoma area. I filed a lawsuit in the United States District Court for the Northern District of Oklahoma and joined all of the defendants in one case. On the same day, the judge granted my motion for a Temporary Restraining Order and seizure order. A deputy marshal and two movers with a truck accompanied me to 14 different locations where we impounded the bootleg SCRAMBLE video games. These locations included a Safeway store in Tulsa, Oral

Roberts University in Tulsa, a 7-Eleven in Biggs, a hamburger stand in Dewar, an arcade in Manford, an arcade in Owasso, and a game room in Bixby.

I took photos for a later press release. One of my photos, showing bootleg video games being taken from the Student Union at Oral Roberts University and being placed on a truck, is shown below. Notice the cowboy hats!

Bootleg SCRAMBLE video games, seized by the US Marshal at Oral Roberts University in Tulsa, Oklahoma, October 1981

During 1981, there were many other seizures of bootleg SCRAMBLE video games. Most of this activity occurred after the appeal hearing at the Court of Appeals. I thought the hearing had gone very well and we were waiting for the decision.

On January 20, 1982, the Court of Appeals issued its opinion, affirming Judge Nickerson on all grounds. We and the entire video game industry were elated. In a landmark decision, the court held that the sights and sounds of video games are subject to copyright and specifically that copyright protection is available for the visual images electronically displayed by a video game. In addition to this case being cited in video game copyright cases, it has also been cited in cases relating to the look of the visual images electronically displayed on a computer screen, even if they result from other than a video game.

Typically, the computer program for the video game is in a PROM. Prior to Omni's appeal, I knew it would be useful to also have the copyright laws apply to the computer program embodied on any type of a memory device, such as a PROM. Then we could take action against pirates who were duplicating my client's PROMS. But at that time, the Copyright Office was not registering copyrights relating to computer programs on a memory device. Copyright attorneys were complaining to no avail.

I traveled to Washington DC and met with top officials of the Copyright Office, including David Ladd who was the Registrar of Copyrights. Dave had been the Commissioner of Patents part of the time when I was a patent examiner. We had become friends in Chicago when he joined the law firm in which I was an associate, although I had not seen him in about 15 years.

The officials at the Copyright Office took the position that if the code was not readable, they were not in position to grant a copyright registration for it. They were willing to grant a registration if the source code were submitted but copyright applicants generally will not submit source code because it is highly confidential. However, we worked out an agreement which is followed to this day, in which submitting a certain amount of an assembly language printout will satisfy the submission requirements for a copyright registration.

After the Copyright Office issued a new regulation based upon our agreement, I obtained numerous copyright registrations for Stern's computer programs that were embodied in PROMs for video games. I also obtained copyright registrations for many other clients outside of the video game industry who used original computer programs. Over the years I filed many lawsuits and obtained numerous Temporary Restraining Orders based upon the copying of my clients' computer programs.

Atari Vs. The Prom Blaster

You are looking at the title of this subchapter and wondering: What is a "Prom Blaster?" The answer is, it is a device that was sold by my client JS&A in 1982 to duplicate Atari video game cartridges. Here's the story.

In 1977, Atari introduced a home video console system for attachment to a TV, the Atari 2600 system. It was also known as the Atari Video Computer System. Instead of the video games being built into the console, Atari sold each video game as a cartridge with the video game program on a PROM inside the cartridge.

Atari 2600 Video Computer System with joystick, circa 1981

Atari licensed some of the most popular video games in addition to producing its own video games, all of which were available to the public by purchasing the Atari cartridges for the games. Other companies, such as Activision, manufactured and sold cartridges for the Atari 2600 system. The cartridges were a tremendous revenue source for Atari and the system was extraordinarily successful.

Under section 117 of the United States copyright laws, if you own a computer program you are entitled to have a backup copy for your personal use to prevent from loss in the event of damage to the original copy. This is referred to as the backup copy exception to the copyright laws. My client JS&A decided to sell a device that would enable purchasers of Atari cartridges to have backups. To this end, JS&A began selling the PROM BLASTER, which duplicated the Atari cartridges.

Top portion of JS&A's PROM BLASTER advertisement

When Atari learned of the PROM BLASTER, Atari filed a lawsuit in federal court in Chicago, requesting a preliminary injunction. Atari claimed that JS&A was selling a device that contributed to the infringement of Atari's copyrights. I now was representing the defendant in a video game copyright lawsuit!

During the hearing on the preliminary injunction, I argued that my client could not be a contributory infringer because the PROM BLASTER's use for providing legal backups was a substantial noninfringing use of the PROM BLASTER. This was the first case in the United States involving the Section 117 backup exception. Atari argued that based on the famous Betamax case, the substantial noninfringing use defense had no merit. This was a difficult issue for me because at the time of the argument, the Betamax video recorder was held by the United States Court of Appeals to be a contributory infringement, although the Betamax case was now before the US Supreme Court for decision.

Based upon the status of the Betamax case at that time, the judge granted Atari the preliminary injunction. I appealed to the United States Court of Appeals for the Federal Circuit. However, before my appeal was decided on the merits, the United States Supreme Court reversed the Betamax case. In the synopsis of the Supreme Court decision the Court stated:

> "The sale of copying equipment, like the sale of other articles of commerce, does not constitute contributory infringement if the product is widely used for legitimate, unobjectionable purposes, or, indeed, is merely capable of substantial noninfringing uses."

After the US Supreme Court decision in the Betamax case Atari recognized that its position was damaged greatly and the lawsuit was dismissed.

For the next almost 40 years, video games have consistently improved and have become extremely popular throughout the world. The Internet has boosted the playing of video games, allowing players who are unknown to each other to compete anywhere. Playing video games over the Internet provides a social connection with friends, family and others who are playing online. As reported in the October 31, 2020 issue of *The Wall Street Journal*, in the United States, video games are played by an estimated 244 million people.

CHAPTER 14

The Battery-Operated Watch

Although this chapter concerns battery-operated watches, I have to admit that I really love mechanical watches. However, battery-operated watches have a strong place in the history of timepieces, and are a technical marvel.

Until I was in high school, battery-operated watches did not exist. Everyone wore a mechanical watch which was energized by manually winding it. Some mechanical watches have an automatic movement so that the turning of the wearer's wrist energizes the mainspring. Mechanical watches have been made for hundreds of years and in many instances are works of art. They are handmade and the movement incorporates gears, barrels and springs to move the hands of the watch. Some mechanical watches can be likened to a beautiful item of jewelry.

The early battery-operated watches did not have a quartz movement but instead used a tuning fork system. The introduction of the quartz movement in 1969 resulted in battery-operated watches becoming inexpensive and very accurate. A mechanical watch movement is typically far more expensive to manufacture then the quartz movement of a battery-operated watch. In addition, a mechanical watch is not nearly as time-accurate as a quartz watch. For example, a fine mechanical watch has an accuracy which varies about five seconds per day. A modern quartz watch may be very close to 100% accurate. Even the cheapest quartz watches will rarely vary more than one second per day.

The first commercial battery-operated watch was the Hamilton 500 series, introduced in 1957. It looked like a mechanical watch but instead of being spring wound, it used a battery-energized moving coil system with drive gears to turn the hands of the watch.

Hamilton Model 505 electric watch, 1957

The next electric watch on the market was highly publicized, and used a completely different method of operation than the Hamilton 500 series. It was the Bulova Accutron watch, introduced in 1960.

Bulova Accutron watch movement, 1961
Photo credit: Rees11 at English Wikipedia. / CC BY-SA/2.5
(https://creativecommons.org/licenses/by-sa/2.5)

The Accutron watch was considered an "electronic" watch, because it used a transistorized (one transistor) oscillator circuit. It was prior to quartz watches and used a vibrating tuning fork to keep time. The photograph above shows the tuning fork prongs at the top, adjacent two electromagnetic coils. The oscillator circuit, shown on the left side of the photograph, energized the coils which drove the vibrating tuning fork so that it vibrated at 360 times per second.

Bulova guaranteed the Accutron watch to be accurate to two seconds per day which at that time was considered incredible. During the 1960s the Accutron tuning fork technology was used by NASA in the space program. However, the tuning fork technology became obsolete in the mid 1970s, once the more accurate quartz watch movement was becoming popular.

The first digital watch that I ever heard about was the Pulsar digital quartz watch, which was introduced (in prototype form) in 1970 on *The Johnny Carson Show*. It was not yet for sale. It was really considered a space-age marvel because it was the first watch that had no moving parts and displayed the time in digits. There were no hands or dial! It's hard to believe but at that time, seeing the hours and minutes is digits instead of as clock hands on a dial was really novel and interesting!

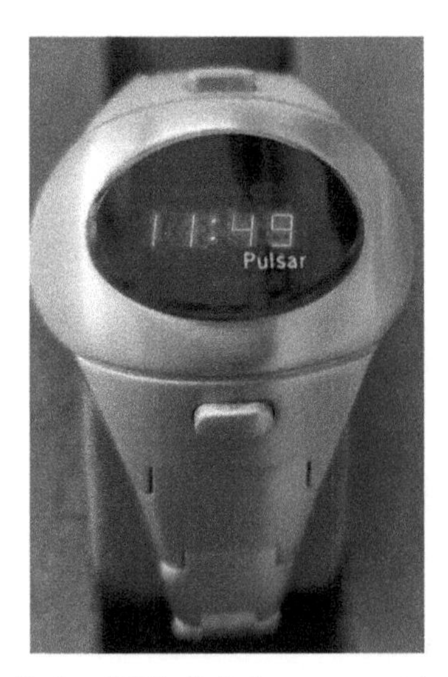

Pulsar LED digital quartz watch
Photo credit: Jan Michael Morris

As shown in the above photograph, the hour and minutes were displayed on light emitting diodes (LEDs). Because LEDs consumed a substantial amount of battery power, the time display would be off until a button was pressed. The time would then light up and if you continued to press the button, the seconds would be displayed.

In 1969 Seiko had introduced an analog-display watch having a quartz crystal movement. This was the first watch in history to have a quartz movement but it did not have a digital display like the Pulsar watch. The quartz movement was an engineering feat and provided far more accurate timing then previous watch movements. The basic timing was provided by a quartz crystal oscillator which was set to oscillate at 32,768 Hz.

The Pulsar watch hit the market in 1972. With a combination of a digital time display and an accurate quartz movement, the Pulsar watch was a sensation at the time. It was accurate to within one minute per year. However, the early Pulsar watches were expensive, with a price tag of $2100. The watch was seen on the wrists of many celebrities.

One main drawback of the Pulsar watch was that the display was totally blank until a button on the case was pressed to reveal the time. Another drawback was the significant battery drain. These drawbacks were overcome later in 1972 with the introduction of the Microma digital quartz watch with a liquid crystal display (LCD). The liquid crystal display drew a miniscule amount of current from the battery and displayed the time with an accuracy within five seconds per month. There was no need to press a button. The time on the LCD display was displayed continuously for the life of the battery, which was approximately one year. The early Microma watches sold for about $300, far less than the Pulsar watch.

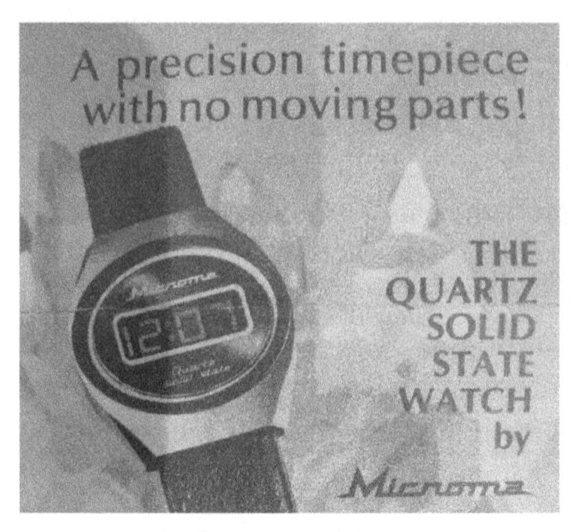

Front page of Microma watch instructions, 1972

The Microma watch used CMOS integrated circuit technology. As stated in the instructions that accompanied the watch, "the Microma watch relies upon a micro-miniaturized computer of amazing capability. On a tiny chip of less than 1/75 of a square inch in area, over 1000 transistors are utilized to translate the precise oscillations of the quartz crystal into an equally precise digital readout of the time."

Microma was a subsidiary of Intel Corporation, the pioneer in integrated circuits. My client JS&A was the first company to sell the Microma watch. I was given one as soon as it was marketed, and I enjoyed wearing it because of its uniqueness and amazing accuracy.

The watch was a real conversation piece. Prior to this watch, no one had ever seen a wrist watch having a constant time display in digits. At restaurants and during other activities, people would come up to me and ask how it works. The art and science relating to watches had taken a new turn in history.

By the mid 70s, digital quartz watches were popping up everywhere and the prices were dropping rapidly. Apparently due to lower pricing and better advertising, the LED watches requiring the pressing of a button were more popular then the LCD watches (with continuous display). By 1977, you could purchase an LED Watch for $10. But then, the popularity of LED watches turned and LCD watches became more popular. Prices of the LCD quartz watches fell and by 1980 you could purchase one for less than $10.

Over the years, the styles and functions changed. Digital watches became multifunctional, with other bells and whistles including, for example, world dates and calculators. The digital display lost its novelty and quartz analog watches, with hands and a dial, became most popular. Many very expensive designer watches including the top Swiss and European watch manufacturers began using both mechanical movements and quartz movements. A large segment of the population preferred a quartz movement with battery power, in contrast to a mechanical movement requiring winding. I understand that more than 97% of watches sold in 2015 had a quartz movement.

At the present, almost everyone in the world is aware of the Apple watch. It is the ultimate battery-operated multifunctional watch.

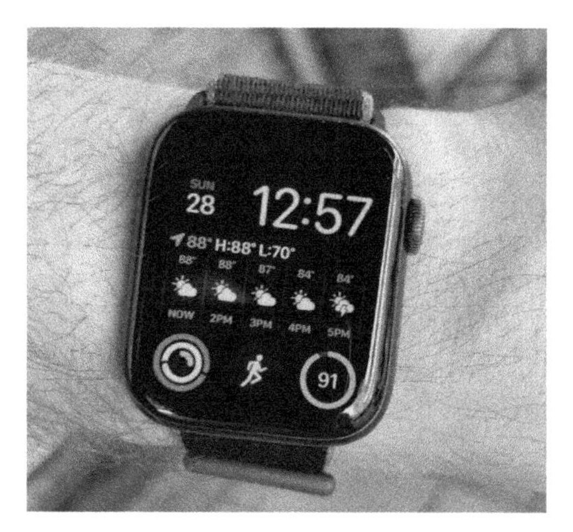

Apple 5 watch, presenting one of its
many digital display modes

The Apple 5 watch has more than 100 watch faces. One of the many faces is shown above, with the digital display. If you want an analog watch display, it can be changed in a moment. The high-resolution OLED display is always on. The time accuracy is perfect. Apple has located 15 network time protocol servers throughout the world. GPS antennas from these servers engage GPS satellites which broadcast time data from the US Naval Observatory in Washington DC, where an atomic clock is located.

The Apple 5 watch provides the wearer with an amazing amount of health data. Electrical and optical heart sensors provide heart rate and electrocardiograms. There are all kinds of exercise and activity regimens. You can customize workouts. You can check noise level. The watch is extremely helpful for traveling, hiking and camping. It uses a Maps app for navigation and has a feature that shows you which direction you are facing. It uses GPS and provides latitude and longitude, and has a barometric altimeter as well as a gyroscope.

The watch contains a speaker and a microphone. You can listen to music on the watch. You can stream the entire Apple music library, podcasts and audiobooks. It has its own app store so you don't need to install watch apps from the iPhone. If you have the cellular option, you do not have to pair with the iPhone and you are able to make and receive telephone calls, and to send and receive email and text messages. You do not need to type your messages; you can dictate them into the watch.

You can see a weather forecast, or use a calculator app, as well as numerous other useful apps. You can surf the web and you can search any topic using Siri. You can use the watch for controlling various electronics remotely. There are many other functions but you certainly get the idea. Apple sells its Apple 5 watch for $399 up, depending upon the casing.

On September 16, 2020, Apple introduced the Apple 6 watch, which includes all of the features of the Apple 5 watch, and, among other things includes a blood oxygen monitor. This feature is particularly significant in view of the coronavirus pandemic of 2020 - 2021, because a low blood oxygen count is a symptom of the virus.

CHAPTER 15

The Office

SOME PATENT OFFICE TECHNOLOGY

My first full-time office employment began in 1960. I was employed as a patent examiner at the US Patent Office located in the Department of Commerce building in Washington DC. The basic duties of a patent examiner were to review patent applications to determine patentability and to issue office actions rejecting, allowing, or objecting to patent claims in the patent application. As patent examiners we were also involved in other aspects of judging patent application matters.

At the time, very few examiners were competent in typing. Ordinarily, the examiner would write the communication (such as an office action) on a paper sheet and place the writing in an in-box to be typed by a secretary in a typing pool. Word processors did not exist. Each secretary had a standard IBM electric typewriter and the writing would be typed into a formal government communication. We tried to write our communications very carefully, recognizing that once they were typed by a secretary, changes may require a complete retyping. Although dictation machines and shorthand were widely used in law offices, they were not used at the Patent Office when I was there.

A standard IBM office typewriter, circa 1960

Nothing was machine copied. In order to have copies, carbon paper was inserted between multiple sheets. For example, five sheets of carbon paper were required in order to produce five copies. When a typing error occurred, the original and all of the copies had to be corrected. Also, each subsequent copy was more difficult to read because there was more paper for the striking typewriter keys to go through. When a letter was sent, the recipients of carbon copies of that letter usually were named after the initials "cc:" which initials meant "carbon copy." A recipient whose name was not intended to be shown to the others was named after the initials "bcc:" which meant "blind carbon copy." To this day, these initials are still being used for recipients of emails and machine copies, even though carbon copies have been obsolete for many years.

One of the duties of a patent examiner is to determine whether the invention claimed in a patent application is disclosed in prior patents or other publications. Now, of course, searching is done by computer. However, in the 1960s through 1963 time period when I was an examiner, there were no computers that examiners could use for searching. There were no available search engines. Searching had to be done manually, by physically viewing prior patents and publications.

The manual search consisted of finding the right search classification, and then looking through references (patents and publications) that were located in "shoes" having that classification. The shoes were small drawers, about the size of patents, containing stacks of references having a particular classification. Examiners would look through the drawings and if something looked pertinent, quickly read some of the text to determine the relevance of the reference. This was far less reliable then a computer search, which did not become available for many years.

Patent examiners searching the "shoes" at the Patent Office, Washington DC, 1963

An example of the unreliability of a manual search through the shoes is as follows. When performing a search, if a relevant reference was found in the shoes, the reference was physically removed from the shoe. It was inserted in the patent application file and that reference, and other relevant references from other shoes, were forwarded to the secretary when the communication was to be typed. The references that were removed from the shoes were not returned to the shoes until the communication was completed and a secretary was available to return them to the shoes. As a result of this procedure, references may be missing from the shoes for more than one week. If another search was performed during the time that the references were missing, that search would be unable to uncover these missing references.

During the three years that I was employed as an examiner at the Patent Office I attended George Washington Law School in the evenings. As soon as I completed law school I was employed as an associate at a law firm in Chicago beginning August 1963. The firm had a long name but I will refer to it as the Dressler firm. The technology in an office such as the law firm in the 1960s as compared with present technology is like night and day.

At the Dressler firm the secretaries had typewriters that were more advanced than the standard IBM electric typewriters. They were IBM Selectric typewriters and they had the ability to handle various fonts, using a removable typing ball ("typeball") instead of a fixed bundle of typing elements. There were different typeballs for different fonts and typeballs with different language symbols and mathematical symbols. Removing and inserting various typeballs was simple.

An IBM Selectric typeball

In the late 1960s, some of the secretaries received new Selectric typewriters with proportional spacing. For example, an "i" would take up less space than an "m." In addition, it enabled right justification. With practice and skill, this resulted in a document appearing to be printed from an expensive printing machine. This was long before the secretaries used word processors.

Advances In Dictation

During my seven years at the Dressler firm and for many years thereafter in the law firm that I founded, every lawyer had a private secretary. Lawyers were not expected to type and all typing and clerical functions, as well as various bookkeeping functions, were assigned to the private secretary. Starting around 2000 the term "secretary" was rarely used and the term used is now "administrative assistant" or "executive assistant." Pardon me but I believe it is proper for me to use the term "secretary" when I am referring to the time period that this job title was used.

Every secretary at the Dressler firm was a highly skilled typist and was able to use shorthand for dictation in addition to using a transcribing machine. Whenever I hired a secretary, I would dictate a letter which she would take down in shorthand and then I would review the letter for accuracy after she typed it. Toward the end of the 20th Century it was much less likely for each lawyer to have a private secretary and as time progressed, each administrative assistant handled clerical and administrative work for several lawyers. Most lawyers now have the computer and typing skills that enable one assistant to handle as many as six lawyers.

Although I enjoyed and became skilled at direct dictation to my secretary, I found that it was often more efficient to use a dictating machine. In this way, I could be dictating into the dictating machine at the same time my secretary was transcribing something else that I had dictated. In both instances it was important to dictate accurately during most of the 1960s, because multiple copies of the documents were made using carbon paper and revisions were difficult. There were no word processors so entire documents occasionally needed complete retyping and changes required skilled retouching.

During my legal career, the technology of dictating machines had a major evolution. When I started at the Dressler firm we used Stenorette reel to reel dictating machines. These were large, cumbersome, and weighed approximately 10 pounds. They would sit on our desks, with a connected microphone. When dictating, you would insert a dictation tape in the tape slot on the left and feed it into the take-up reel on the right side. You would then press the "Dictate" button. When you completed dictation on the tape, you would rewind the dictation tape and give it to your secretary for transcribing. She would insert the tape in her machine, press the "Listen" button and use a connected foot pedal for advancing the tape.

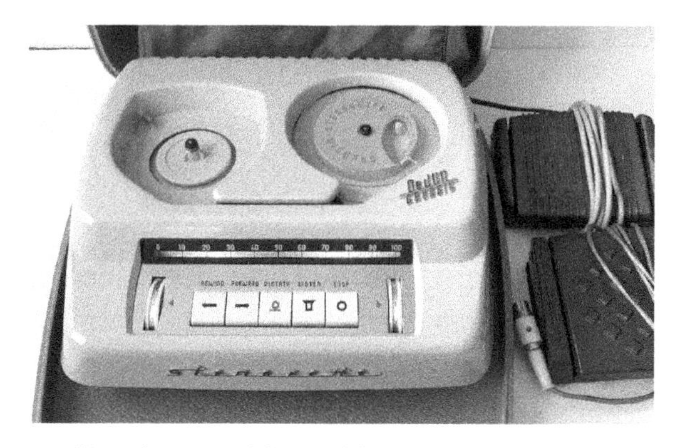

Stenorette dictating machine with two transcribing foot pedals
(microphone not shown), circa 1967

In the 1970s, the size of dictating machines greatly decreased. With the advent of tape cassettes and integrated circuits, dictating machines were introduced that were battery-operated and portable. They were usually called "voice recorders." Now I was able to go from a cumbersome ten pound machine to a hand-held battery-operated voice recorder that I could carry in my pocket. These voice recorders had built-in microphones and speakers, and finger controls for dictating, stopping, reversing, and erasing.

It seems like I was using my hand-held Dictaphone voice recorder everywhere I went as well as in the office. I did very little written notetaking: I found it easier to dictate whatever notes I needed. In fact, I dictated almost everything. When my dictation was completed,

I would remove the mini-cassette from the recorder and give it to my secretary for transcription. She had a transcribing machine that handled mini-cassettes. I used this mode of dictation throughout my legal career.

Dictaphone Model 3232 hand-held portable voice recorder, circa 1975

The problem that I had with the Dictaphone voice recorder was that my constant manipulation of the finger controls resulted in the recorder needing frequent repair. The Dictaphone recorder and the after-warranty repairs were quite expensive so I found it more efficient to also use inexpensive Sony mini-cassette voice recorders and discard them when they needed repair. Also, I kept several of them in various different places so there was always one available for me to use.

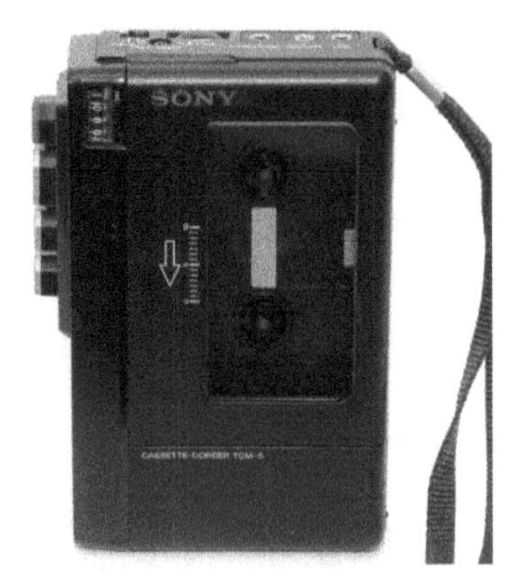

Sony Model TCM-5 mini-cassette recorder, circa 1990

A more recent type of dictating machine is the digital voice recorder. The digital voice recorder does not use analog audio tape like the Dictaphone or Sony recorders pictured above. Instead, the voice recording is digitally recorded on internal storage media or on a memory card. After the dictation is completed, the digital data is transmitted to an assistant for transcribing without involving any audio tapes.

Some systems used in an office utilize docking stations in which the hand-held recorder is positioned to transmit the data directly to a transcriber in the office. Many digital voice recording systems can send the digital data to the transcribing assistant via email or using a dedicated app on the computer. In addition, there are dictation apps for smartphones such as the iPhone, in which you can dictate and then send the dictated data via email or to an associated app at the transcribing station. You can see that digital voice recording enables dictation to transcription over the Internet anywhere the world without requiring the conveyance of any audio tape or other physical object.

What do you do if you have no typing skills, are in an office without an assistant to help you, and you need to produce typed documents? The solution is called "speech to text." I have used it constantly since I retired from law practice.

My first introduction to speech to text was when a sales person came to my office in the 1990s with a computer and a microphone system connected to the computer. He dictated into the microphone and his words were shown on the computer display. He had dictated very slowly and it appeared to be about 90% accurate. I tried it, dictating slowly and clearly, but found it to be extremely inaccurate. I was told that the system would have to learn my speech pattern. I really had no interest in this system because of its difficulty and, in any event, I had an excellent private secretary.

Shortly after 2000 I was at a conference where they were demonstrating a speech to text system called Dragon NaturallySpeaking. I tried it out and found it to be more accurate than the system I had used previously. However, it still seemed to be inaccurate and I had no real interest. On the other hand, not being a typist, the idea intrigued me.

In 2011 I decided to write a book about my professional life. I also decided that speech to text would be perfect for me if I could master the technology with accuracy. I purchased a Dragon NaturallySpeaking program for my Dell Computer and experimented with it. After a couple of days, I was pleased with the results. The manuscript for my book, which was published in 2013 with the title *Clear and Convincing Evidence*, was typed with my dictation and edits. All 355 pages with more than 100,000 words!

All of Apple's IOS devices now enable speech to text dictation. The latter portion of my *Clear and Convincing Evidence* manuscript was dictated directly into my iPhone using the iPhone dictation setting. Now I dictate almost everything that would normally require a keyboard. This includes emails, notes, text messages, and even most of the manuscript for this book. It's so simple on the iPhone. Whenever the keyboard pops up for typing, simply hit the microphone symbol and dictate the text instead of typing it.

iPhone X keyboard, with
dictation microphone at bottom right

The Xerox Copy Revolution

When I was in elementary school, high school and college there was no such thing as Xerox copies. There was no Xerox copy machine. Even when I began my career at the Dressler firm in the early 1960s, the firm did not have a Xerox copy machine. We had a Kodak Verifax copier, which was probably the most popular copier in the United States at the time.

Kodak Verifax Bantam copier (closed) circa 1960
Photo credit: Ladonna

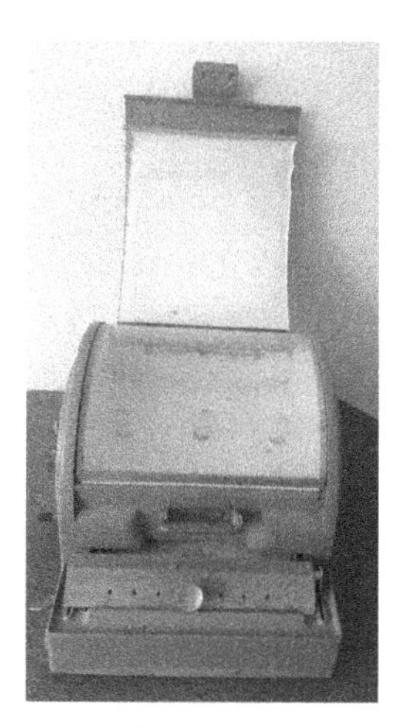

Kodak Verifax Bantam copier (top open for copying) circa 1960
Photo credit: Ladonna

The Verifax copier, as well as other copiers on the market, were extremely slow and produced low quality copies. For that reason, copies of documents were made infrequently and only when absolutely necessary. Numerous steps are required to make a copy. You first took a photosensitive matrix paper and placed the matte side of the matrix paper on the curved glass. The curved glass is shown on the second photograph with the lamps underneath it. You then placed the original over the shiny side of the matrix paper. Then you closed and latched the platen as shown in the first photograph. You then exposed the papers to light from the lamps underneath the glass.

The next step was to remove the matrix paper and insert it into an activating tray containing a developing solution. You would then remove the wet, developed matrix paper from the tray and it was a negative of the desired copy. You then took a sheet of sensitized copy paper and pressed the wet matrix paper against the copy paper using a squeegee roller in the copier. The copy paper then had to be separated from the matrix paper, resulting in a mediocre copy on a slightly damp sheet of copy paper having a chemical smell. You now had one copy! After the copy dried, over a period of time it would lose contrast (between the paper and the words). I had some old Verifax copies inside of my early legal files from the Dressler firm. After five years many of them were unreadable.

The Xerox copier changed everything. Now you could easily obtain multiple extremely high-quality copies on plain paper. During the early 1960s the world was introduced to what was commonly referred to as a "Xerox Machine."

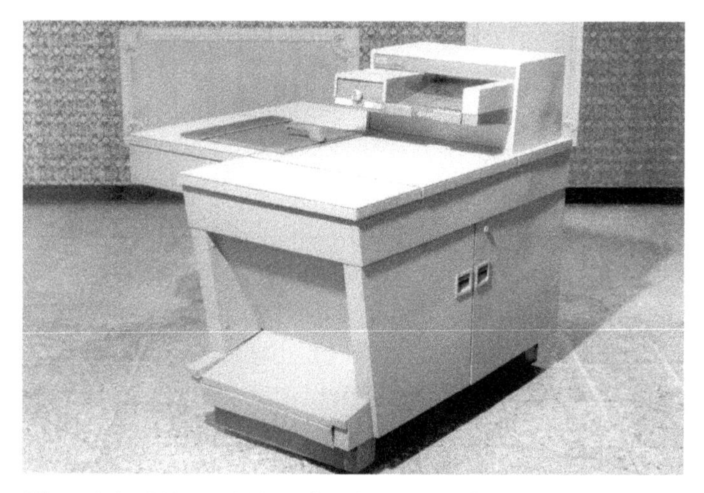

The original Xerox 914 copier. Courtesy of Xerox Corporation

The first Xerox machine that I ever saw was in 1963 at the Department of Commerce building in Washington DC, where the US Patent Office was located. I'll never forget how amazed I was to see a high quality copy on plain paper. As a patent examiner, I needed to make a copy of a reference page comprising an extremely detailed electrical circuit. The lines of the circuit were very close together and the letters and numbers were extremely small, so a Verifax copier could not handle it properly. The Department of Commerce had a new Xerox Model 914 machine and I remember making two or three copies of the circuit diagram on the Xerox machine. The most amazing thing was that I could not tell the original from the copies! I remember bringing the original and a copy home to show my wife how a copy on plain paper looked exactly like the original.

The reason why Xerox gave its first machine model number 914 was because it was capable of making copies up to 9 inch X 14 inch. It was the largest and most expensive copy machine that I had ever seen. It weighed about 650 pounds and cost the government $26,500 (in 1963 dollars). It was so easy to use. You would simply place the original face down on the glass plate, indicate the number of copies you want, and press a button. The ordinary paper that you had previously stacked in the copier would be used to provide the copies. This is like copies are presently made but quite a difference from complex photochemical copying.

Although during my first two years at the Dressler firm we used an inexpensive Kodak Verifax copier, in 1965 the firm obtained a Xerox 914 copier. Xerox had made it much more available to businesses by leasing it for $25 a month plus $.10 per copy. The Xerox machine completely changed the manner in which copies were made at the Dressler firm. Instead of always making carbon copies of typed documents, an original document was typed and the copies were made using the Xerox machine. The clients were charged for the copies and no one ever complained about it.

The Xerox machine was ideal for making copies of patent applications. At the time, they were required to be on legal size paper with a particular format. Many were very lengthy and numerous copies had to be made using carbon paper. With the Xerox machine, an original document would be typed and the copies would be made using the Xerox machine. In addition to resulting in very clean copies similar to the original, there was no need to make corrections on carbon copies anymore. Interestingly, the inventor of the Xerox machine, Chester Carlson, was a patent attorney working for Bell Labs. One of his reasons for inventing the Xerox machine was to make life easier in typing patent applications!

The Xerox machine was one of the most important developments in the 1960s. It became a "must-have" in most businesses. If you watched the TV series *Mad Men*, you probably remember when the Xerox 914 copier was delivered to the office. It was stationed in a place of great importance and received a grand reception. The term Xerox became so attached to copying that instead of referring to making copies of document as "copying" it was often referred to as "Xeroxing."

By the end of the 1960s, Xerox machines and competitive copiers were found in almost every office, and carbon paper became obsolete. The new copiers were becoming faster and cheaper and color copying became available. The technology was developed so that copiers were also operable as printers for computers. With the explosive use of personal computers, home printer/copiers were developed that were extremely inexpensive, with the printer/copier manufacturers making up for the low cost of the printer/copiers with the price of the required toner cartridges. By 2000, small multifunction units, with the ability to print in black-and-white or color, scan and fax, became the norm in home offices.

The Fax Machine

One of the most welcome devices in the office was the fax machine.

At present, technology has made communicating so simple. All types of communication, written, printed, photography, videos, etc. can be sent and received instantaneously. This is enabled through the marvel of email, text messages, and hundreds of apps including Facebook, Instagram, LinkedIn, Twitter, and so many others.

If we want someone anywhere in the world to see our communication, all it takes is a couple of strokes and the communication is on its way. This is not at all how it was in 1960, when I started in the business world. At that time, there was no Internet. Even Federal Express wasn't in business until 1973. if we wanted to communicate other than by voice over the telephone we had to use the mail or a personal courier.

As an example, if I wanted to file a patent application in Germany, I would have to send the papers by international mail to a patent agent in Germany. It wouldn't be unusual for the papers to take two weeks to reach Germany. The German agent would have to send the translated papers back to me which could take another two weeks. The returned signed papers could take another two weeks. Likewise, amendments in the file would take these amounts of time to travel back-and-forth between Germany and the United States. Large documents were usually sent via regular international mail, which meant traveling by ship across the ocean. Communications in the form of short letters could be sent via air mail, traveling by airplane instead of via ship, which would normally take only a few days instead of a couple of weeks. But everything was far from instantaneous and the delays were more than inconvenient. Consider an international contract that would have to go back-and-forth to be revised and signed. Even using air mail could take weeks by the time the contract was revised several times.

Even though Federal Express came into existence in 1973, it did not provide instantaneous communication – – it only provided overnight communication. This was helpful compared to the United States mail but it was hard to realize how important instantaneous communication was until the advent of fax machines.

The fax machine enabled the contents of a document to be transmitted instantaneously to a receiving party anywhere in the world via a telephone line. In order for this to be accomplished, you would need a telephone line and a fax machine comprising a scanner and a printer. The earlier fax machines incorporated a thermal printer but in the 1990s, plain paper printers were introduced. In addition, software was introduced that would enable faxes to be sent directly to your PC.

The basic operation of a fax machine is not complicated. The fax machine contains a scanner which will scan the document to be sent. The black and white information resulting from the scan is converted into a sound frequency signal. The sound frequency signal is transmitted over a conventional telephone line to a receiving fax machine. The receiving fax machine will convert the sound frequency signal into black and white information which can be printed to become a facsimile of the original document that was transmitted.

The scanner that obtains the black-and-white information uses a CCD (charge coupled device) sensor which converts the analog pattern of black-and-white areas of the document into a digital pattern of binary 1's and 0's to form the information that is sent over the telephone line.

The fax machines in the 1980's required a dedicated telephone line and were relatively expensive. My office at first used a local fax service and later used what is called an eFax. This was a service that provided us with a dedicated phone number and all faxes sent to that phone number would be forwarded to us using that service via email. Subsequently we purchased a Brother fax machine and leased a dedicated fax phone line.

Brother IntelliFAX plain paper fax machine

A problem with the fax machine is that document transmission is very slow because of the slow transmission rate of a telephone line. It's like using a dial-up modem. A document containing 30 pages may take 30 minutes to transmit. A detailed document will take longer to transmit than a simple, mainly blank document because more information has to be conveyed. Noise in the telephone line may also cause a transmission problem.

Further, the quality of the faxed document is relatively poor because the scanning operation does not handle all of the grays. Too much transmission speed would be needed for that much information. Telephone modems transmit only up to 56K bits/second while modern broadband modems transmit in megabits/second. Documents containing greys (e.g. photographs) and having colors other than black-and-white are transmitted with much greater quality by email or other programs over the Internet using a broadband modem instead of a telephone line as used by fax machines. However, fax machines provide satisfactory transmission of simple documents with black type on white paper. In addition, because of certain perceived security concerns, fax machines are still used in medical facilities and in governmental offices.

Although fax machines seemed to be a new concept in the 1980s, they really weren't. In the mid 20th century, fax machines, although unheard of by most of the public, were used by Western Union. The technology was different from today's fax machines. Many of the telegram delivery cars used by Western Union carried a fax machine in the car with the power source and other vacuum-tube electronics in the trunk. The central office of Western Union would receive a message to be delivered and the message would be wrapped on a roller in the central office fax machine, scanned, and transmitted to the fax machine in the car. The driver would receive the fax on conductive paper wrapped on a roller and personally deliver it to the recipient. These fax machines were manufactured by Seeburg (the juke-box company) for Western Union, and Western Union found a market for them in government offices. The size of the fax machine was reduced and it was known as the Western Union Desk Fax, some of which are now collector items.

Western Union Desk Fax, circa 1948
Photo credit: retroversepnw (ebay)

The fax machine, even if used infrequently, has now become an integral part of almost every home/office multifunction copier, with the concomitant benefit of the use of plain paper for printing.

CHAPTER 16

Radio Days

As I discussed earlier in this book, prior to the introduction of television we were entertained by radio programs. There were no video games or personal computers. Listening to the sounds from the radio and using our imagination with respect to visualization was a pleasure. The daily serial programs on the radio, with the sound effects, seemed real and it was fun to imagine what was happening. The daily serial drama and action programs are practically nonexistent today, but listening to music over the radio has persisted. One area of entertainment that has really mushroomed on the radio is "talk radio."

Radio design was an important art form. Until the 1950s the radio was a significant piece of furniture in the home. Tabletop radios were popular, and homes throughout New York City had large, console radios in the living rooms. When we moved from Manhattan to Forest Hills, Queens, New York in 1941, my parents brought along a Stromberg-Carlson console radio. We kept it in the living room until 1950, when it was replaced by a television set. It was on constantly.

If you looked at the back of the radio, you would see the vacuum tube circuitry sitting on a shelf and the single very large speaker below the shelf. This was conventional design for console radios. The radio handled the AM broadcast band and used a tuning dial for finding a station. At that time there was no digital scanning.

Stromberg-Carlson Model 140-L console radio, circa 1937

There was also a phonograph socket on the back of the radio. We had a 78 RPM turntable that we would plug into the phonograph socket and it would use the amplifier and speaker of the radio. My parents had a large collection of records and my brother and I would sit in the living room, listening to records being played. One that I specifically recall being played often was the orchestration of "Peter and the Wolf."

My brother and I play a 78 RPM record, circa 1942

About 1949 we obtained an RCA Victor 45 RPM record player/ changer. This was a very popular item, dedicated to playing 45 RPM records and was able to stack up to twelve RPM records. We would plug this unit into the phonograph socket of our Stromberg-Carlson radio in order to use the amplifier and speaker of the radio.

RCA Victor Model 45-J-2 45 RPM record player

Although I did not appreciate it at the time, many of the designs of the vintage radios were beautiful, turning these radios into collector items. My favorite radio design is Art Deco, which was used competitively throughout the 1930s in particular. One radio that I collected that I found very attractive was the Majestic Model 773 “Lido” which was introduced in 1933.

Majestic “Lido” console radio

Just like in our Stromberg-Carlson console radio, the vacuum tube circuitry of the Majestic Lido radio sat on a shelf with a single large (15 inch) speaker located below the shelf. A working vintage radio is presently worth far more than one that is not operative. There are many collectors that are handy in restoring vintage radios, and the vacuum tubes can usually be found on eBay and at many other sources.

Rear view of Majestic "Lido" console radio

In the late 1940s my brother and I had a tabletop radio in the room that we shared. These radios used vacuum tube circuitry and a much smaller speaker than used by the console radios. They were relatively inexpensive and were found in most homes throughout the United States. We had a simple AM band radio and to us, it had everything that we wanted to hear. The AM radios had all of the stations with the drama and action serials as well as top popular music.

RCA tabletop Bakelite AM band radio, circa 1947

Also, in the late 1940s we obtained a portable AM broadcast radio that we used to take to the beach on Long Island. Portable radios at that time were similar in width and length to tabletop radios but were thinner. They usually weighed at least five pounds with a 45 volt radio battery.

RCA portable AM radio Model 8BX6 "Globetrotter"

Open rear view of the RCA "Globetrotter" portable radio
showing the vacuum tube circuitry, the speaker
and the lower battery compartment

Although we now are using Internet radio and satellite radios in cars, in the 1940s and 1950s car radios relied on vacuum tube circuitry and were relatively heavy and expensive. Until the 1950s, car radios only used the AM broadcast band.

1955 Ford AM car radio with push-button presets
Photo credit: Rychy Car Audio and More LLC

Rear perspective view of the 1955 Ford AM car radio
Photo credit: Rychy Car Audio and More LLC

A car radio was considered an optional accessory. If a car being sold had a radio, it was specifically pointed out in the advertisement that it had the radio. Until about 1960, even the heater was an optional accessory! If a used car being offered for sale in a classified advertisement had a radio and heater it would be specified or you would typically see the designation “r & h.” Even electric turn signals were optional until about 1960. If you didn’t have electric turn signals the driver would be required to extend his or her arm out the window to signal a turn.

I have fond memories of listening to the radio during certain car trips. During the 1950's, when we would drive at night we would often listen to country music. I remember listening to Johnny Cash, Tammy Wynette, Hank Snow, Kitty Wells, Patsy Cline, and many others. We also listened to serials such as *Amos and Andy.* When driving during the daytime, the radio was typically tuned to popular music, and we listened to vocalists such as Johnny Mathis, Nat King Cole, Eddie Fisher, Elvis Presley, The Platters, Little Richard, The Everly Brothers, Fats Domino, and many others.

The problem with AM/FM radio is that its radius of service was very limited. For example, an AM radio station could broadcast for only about 100 miles during daylight and perhaps 500 miles at nighttime. FM radio stations were even more limited and only broadcast up to about 40 miles. In order for the radio stations to extend their coverage, they needed additional radio transmission towers. Typically, as you would travel by car listening to the radio the station may fade away and you would have to select a different program on a different station. This was very different from the present use of car radio, in which almost all cars are built with free AM/FM and sound systems incorporating satellite radio (for a subscription fee) and/or Internet streaming radio.

The most significant change in car radio occurred about 2005, when radio began being transmitted to consumers via satellites over 20,000 miles away in space. The signal transmission range covers thousands of miles over the earth. The radio signal from the satellite is scrambled and sent with metadata. The car's receiver unscrambles the data and information about the broadcast is displayed on a screen within the car.

Sirius XM, a satellite radio broadcaster, has a deal with most car manufacturers. When purchased, the car comes equipped with the Sirius XM equipment and a three-month free trial. Before the end of the three-month trial, the car owner is urged by email and regular mail to purchase the Sirius XM monthly subscription service.

Sirius XM satellite radio on the display screen of a 2020 Audi

One product that is sometimes considered to have made the greatest change to consumer electronics is the transistor radio. I remember the first time I saw a transistor radio. I was in my Forest Hills High School junior year home room class in 1955. One of the students brought to the classroom his new Zenith Royal 500 transistor radio. The whole class gathered around as he demonstrated it. Everyone was fascinated by the small size of the radio (5.5 inch X 3.5 inch X 1.5 inch). It was small enough to fit in a pocket and weighed only a pound.

Zenith Royal 500 transistor radio, 1955

The Zenith Royal 500 radio had seven transistors and used four AA batteries, in sharp contrast to portable vacuum tube radios using a large and heavy 45 volt radio battery. It was tough, made of nylon, and the solid-state electronics overcame the fragility of vacuum tubes. At a retail price of $75 in 1955, it was considered very expensive. It only handled the AM broadcast band but the sound quality was very good.

The technology of transistor radios has evolved into all types of recording and playing devices that we can easily carry with us. We now can listen to AM radio, FM radio, satellite radio and Internet radio over our smartphones, thanks to the development of the transistor and micro-miniaturization.

CHAPTER 17

Pinball

I've always loved playing pinball. I even enjoy watching pinball being played by others. I appreciate the artistry of the game playfield and backbox and the action involved in order to obtain the highest score. Unfortunately, when I was younger, between the 1940s and the 1970s, coin-operated pinball machines were banned in many of the large cities, such as New York, Chicago and Los Angeles. Apparently, those city governments considered pinball to be a type of gambling. Also, certain of the distributors were considered to be "mob" connected.

In the photograph below taken in 1949, NY Police Commissioner William O'Brian is shown destroying a pinball machine that had been confiscated on a raid by the police.

Photo credit: Bettmann/ via Getty Images

During the time that pinball machines were banned inside of the large cities, I found pinball machines in areas outside of those cities in general stores, taverns, restaurants and train stations. In the late 1940s and during the 1950s, pinball machines had wood cabinets and wood playfields. As is still conventional, the coin slot and plunger were at the front end of the cabinet. It would cost only a nickel for three to five balls, although later in the 1950s the price was increased to a dime.

The term "pinball" was particularly applicable to the pinball games of the 1920s and 1930s, which were not electro-mechanically operated. They had an inclined (roll-down) playfield with numerous holes and pins (nails) that were strategically located. The plunger would send steel balls into the playfield and they would rebound off pins and land in holes, each of which had a particular score amount. Balls missing the holes would flow into a drain at the bottom of the inclined playfield. You can view online the play of a 1930 pinball machine from a collector's YouTube video at www.gerstman.com/vintagepinball.htm.

There is something very nostalgic about the old electromechanical wooden pinball machines of the 1940s 1950s and 1960s. Maybe they bring back memories of the summertime when I was out of town and would play the pinball machines at a small-town store or tavern.

"Singapore" electro-mechanical pinball machine
made by United Mfg. Co., circa 1948

In the 1940s through the 1960s pinball was extremely popular and there were dozens of companies manufacturing them, mostly located in Chicago. The Singapore pinball machine shown in the above photograph is similar to many that I played. It was introduced in 1947 and had a coin slot at the front for receiving nickels (or dimes) and a plunger to the right for propelling the balls. The Singapore pinball machine was not manufactured with flippers because they had just been introduced by Gottlieb Manufacturing in 1947, although flippers were subsequently retrofitted onto many of the pinball machines of the 1940s.

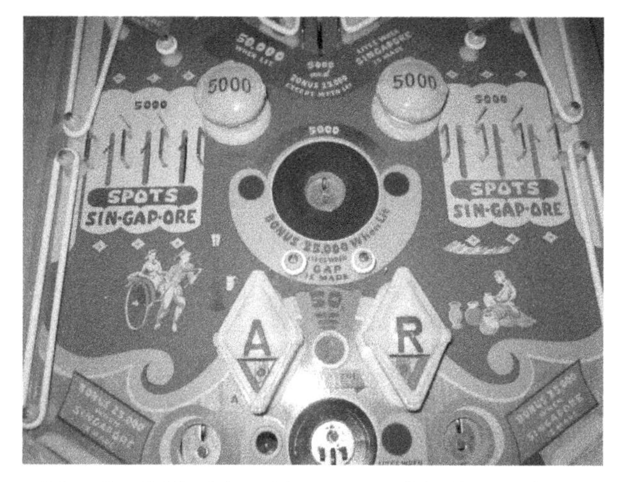

Portion of the playfield of the "Singapore" pinball machine, circa 1948

Even now as I look upon the photograph showing the playfield of the Singapore pinball machine, I can recall how we tried to score the most points possible by propelling a ball into the playfield. There were bumpers that increased the score by 5000 each time they were hit. There was a kick-out hole that would increase the score by 5000 points when the ball rolled in. However, if lit it would provide a 25,000 point bonus. You would also try to get more points by hitting bumpers and going through sliders that had letters making up the word "Singapore." In addition to an increase in score, every time the ball hit an obstacle or went through a designated pass the machine would provide special sounds and lights.

Most of the machines would award extra balls if certain scores were achieved. Some of the machines would award free games in response to especially high scores. The pinball machines with flippers provided players who were skillful in using the flippers an additional source of high-scoring. The flippers were often located near the drain and by using the flippers strategically with good reflexes, a player could propel the ball back up into the playfield and prevent the ball from going down the drain. While playing the game, it was also helpful to nudge the machine slightly in order to urge the ball toward a certain bumper or to place it into a desired alley. However, all of the pinball machines had tilt mechanisms so this had to be done very carefully in order to avoid the tilt mechanism from terminating the game.

The tilt mechanism was far from perfect. There was a method in which a player could consistently achieve a high score, enabling him to receive extra balls or free new games. Before starting the game, the player would carefully place the two front legs of the pinball machine on top of his toes. If this was done carefully enough, the tilt mechanism would not be actuated. This could be accomplished with regular shoes but not with sneakers. When the front legs of the machine were on the player's toes, the playfield was no longer inclined downward toward the drain but instead was substantially level. In this manner, the ball was kept on the playfield for a long time instead of flowing downwardly via gravity. The primary thing that would bring the ball downward would be the recoil from a kick-out bumper or the like that would send the ball careening down to the drain.

In 1976 a client of mine, Universal Research Laboratories ("URL"), invented a different type of pinball machine that was unique. At the time, air hockey was an extremely popular game. URL's new invention was a combination of an air hockey game and a pinball machine. Instead of using balls, it used plastic pucks which moved over a perforated playfield that operated like an air hockey game. A fan circulated air under the playfield to enable the puck to float around rapidly as it hit bumpers and flippers, scoring points and making sounds.

URL built a prototype which worked satisfactorily. However, to me it was not nearly as much fun as playing regular pinball using steel balls that engaged all kinds of obstacles and paths.

In 1978 URL was purchased by Stern Electronics, a pinball machine manufacturer, and Stern abandoned the project. Stern also manufactured video games and Seeburg jukeboxes. I had filed a patent application on behalf of URL with respect to the pinball machine. Stern Electronics became a client of mine and after abandoning the project, they assigned the invention to me. The patent subsequently issued as US patent 4,173,341, but the invention was never commercialized.

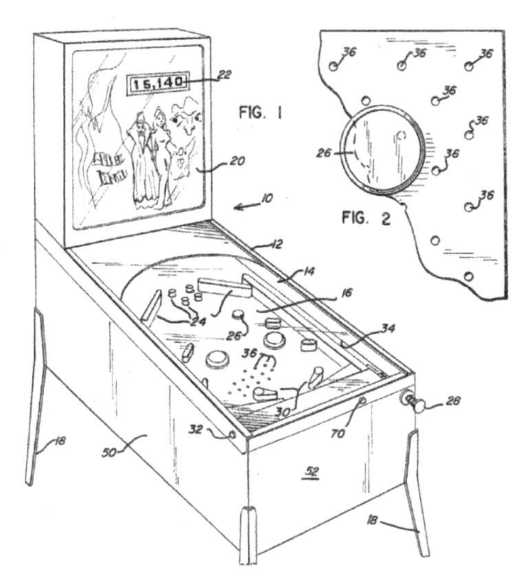

Figs. 1 and 2 of US Patent 4,173,341 for a pinball machine using the air hockey principle, showing puck 26 on a perforated playfield

I really enjoyed handling intellectual property legal work for Stern Electronics, which was one of the most important coin-operated amusement device companies in United States. They manufactured three of my favorite toys: pinball machines, video games, and jukeboxes, and I had the opportunity to learn a great amount about all of them. The company was run by Gary Stern, who was a legend in the business. Pinball machines are very difficult to manufacture because of the large amount of parts that have to be connected by hand. Gary created an assembly line that was amazingly efficient and extremely interesting to watch in action.

In the 1970s and 1980s I was involved in a substantial amount of litigation for Stern Electronics, mainly for videogames but also in connection with pinball machines. This was the era of microprocessor-based electronic coin-operated machines, and the law was being developed. One of the cases that I handled for Stern Electronics (*Stern Electronics v. Kaufman*) was the landmark case deciding the copyrightability of the images and sounds of a video game.

Beginning in the late 1970s, pinball machines became computerized, using microprocessor control. This enabled the machines to have playfields replete with complex electronic targets and obstacles, spectacular sounds and video, and programmed themes of play. The major pinball machine and video game manufacturers, including Stern Electronics, would introduce theme-based pinball machines and video games. For example, if a movie such as *Back to the Future* was about to open, Stern would obtain a license to introduce a pinball machine and/or a video game with the theme and name of *Back to the Future*. The idea was to introduce the game at the same time that the movie would open. This required that the pinball machine and/or the videogame be designed and manufactured on an extremely short time basis.

Just when pinball machines and video games were at their height of popularity, around 1981, the Federal Communications Commission (FCC) proposed a new rule that would slow down production significantly. There had been some complaints from police departments and fire departments that pinball machines and video games and other electronic items using microprocessors were creating electromagnetic interference with their communications. The FCC proposed rule, called "Part 15," required every amusement device operating with a microprocessor to be constructed or shielded to prevent electromagnetic interference and to be tested by an FCC certification official in advance of shipping. Gary Stern immediately recognized that this rule was untenable. Requiring every machine to be certified in advance by an official would cost weeks if not months of delay.

I got in touch with Newton Minow, who was the FCC Commissioner under President John F. Kennedy. He was famous for his statement in 1961 that television programming is a "vast wasteland." He referred me to one of his law partners, Lee Mitchell, who was an FCC expert. There were meetings with Lee, me, and leaders of the amusement device industry, where we discussed methods of alleviating this problem. We came up with the concept of self-certification. Each machine would be tested at the end of the production line and a certification label that it complies with Part 15 would be attached. Lee filed a brief with the FCC, met with FCC officials, and it was accepted. Essentially the same rule is presently in effect. The industry was saved!

Stern Electronics and its successors Data East Pinball and Stern Pinball became the major forces of the pinball industry throughout the world. When the pinball industry collapsed around 2000, Stern Pinball continued to manufacture pinball machines. Stern Pinball became the only major pinball machine manufacturer in the world. There was essentially no competition! There is an art and science in designing pinball machines and Stern had the most talented designers combined with Gary Stern's unbelievably efficient assembly line. This enabled Stern to produce the most complex team-based pinball machines that kept players around the world engaged. The game no longer cost a nickel or a dime to play— it now cost a dollar a game.

Stern Pinball "Star Wars" pinball machine playfield, 2018

Modern pinball machines are far more complex than the pinball machines of the 1940s through the 1960s. They now have superior sound systems, intricate theme-based play, and back box theme-based video. The *Star Wars* pinball machine pictured above was designed by Steve Ritchie, a legend in the field who designed many of the top-selling pinball machines manufactured by Williams and then by Stern Pinball. In addition to having numerous bumpers, kickers and targets, there are rollover switches, ramps, gates, and toys including a ball that runs down a ramp. You can view online the *Star Wars* pinball machine in action on YouTube at www.gerstman.com/starwars.htm.

Many of the pinball machines have "multiball," in which more than one ball is active at a time. There's no limit to the type of devices that can be activated by the ball rolling over the playfield. Everything is controlled by the appropriate software.

One of the top designers whom I worked with at Stern was Joe Kaminkow, vice president of engineering. Between 1990 and 1998, my law firm obtained 17 patents on pinball machines in which Joe was an inventor. He designed theme-based video games as well as theme-based pinball machines and was known as a top creator throughout the industry. In 1999 he left Stern Pinball to join International Game Technology (IGT), the largest slot machine manufacturer. He revolutionized the theme-based slot machine, which is now a staple in the industry. Joe was a master at obtaining licenses from movie studios to use the name and the theme of the movie in connection with the slot machine. He had brought with him his experience in dealing with such licensing in connection with video games and pinball machines. IGT's stock multiplied greatly as a result of Joe's creative endeavors. IGT, located in Reno, Nevada, was a wonderful client of mine and it was exciting to watch the new slot machine technology emerging.

The theme-based slot machine would not be in existence but for the technological change in the late 1970s from mechanical slot machines (with rotating wheels) to electronic slot machines. Coincidentally, the electronic slot machine was invented in 1974 by Bill Olliges, president of my client Universal Research Laboratories. Bill's electronic slot machine was the first slot machine to utilize symbols displayed on a video display screen (at that time, a cathode ray tube). A randomizer circuit was utilized in the slot machine under the control of a microprocessor-based electronic circuit.

I filed a patent application on the invention in 1974. While the patent application was pending, the invention and application were purchased by Bally Manufacturing Corporation and Bally took over the prosecution of the application. The patent issued under US patent no. 4,648,600 and is entitled "Video Slot Machine."

The entire slot machine industry and the casinos changed over from mechanical slot machines to electronic slot machines.

Mechanical slot machines are now collector items. There were many patent lawsuits over these new electronic slot machines and one of them involving the Olliges patent was brought by Bally against my client IGT. Because I was the person who obtained the original information for preparing the patent application and I was the person who prepared the patent application, IGT took my deposition during the lawsuit. As a result of the lawsuit and other relevant patents, there is a substantial amount of patent cross-licensing between the various slot machine companies.

The bridge between pinball machines and slot machines occurred because of the advances in technology. Primarily mechanical devices have been transformed to electronic devices as the result of transistors and microprocessor-based circuits. Computer programming provides the operational instructions. We now see video presentations on the pinball and slot machine displays. Aren't the present pinball machines and slot machines actually forms of video games?

CHAPTER 18

Buying Things

CREDIT CARDS

Over the years, I've seen an unbelievable evolution in methods of shopping. In the 1940s and 50s, you would use cash, checks or money orders to purchase everything. There were no credit or debit cards that could be used in a supermarket or in more than 99% of the stores throughout the United States. Certain department stores, such as Bloomingdale's, provided and accepted "charge plates" that were small metal plates looking something like military dog tags. They were embossed with the customer's name and address.

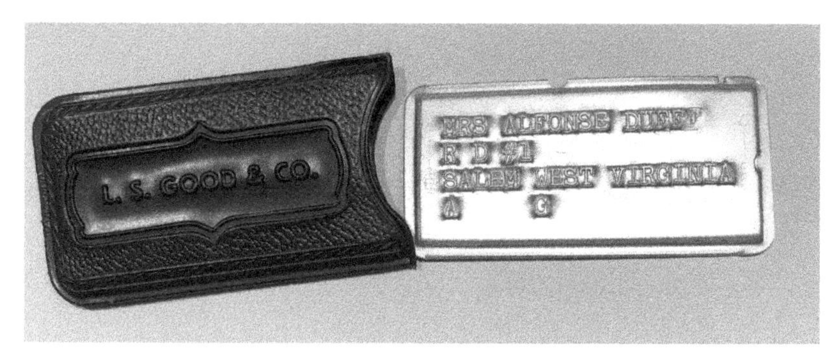

A charge plate and holder from L. S. Good & Co., circa 1955

These charge plates were only about one-third of the size of present plastic credit cards. A paper signature card was held on the back of the plate.

Back of charge plate and holder

When a purchase was made, the charge plate was placed into an imprinter that used an inked ribbon to make an impression of the embossed information on a charge slip. The charge slip was used for bookkeeping: there was no connection to a computer or any other electronic device. My mother had a Bloomingdale's charge plate, and this was the only knowledge that I had at the time regarding charging a purchase.

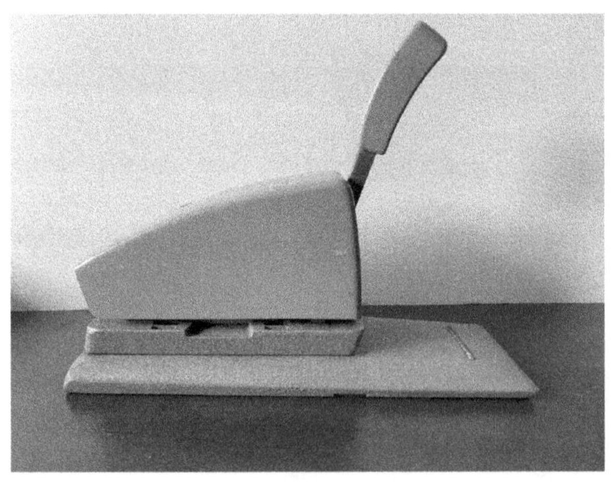

Farrington charge plate imprinter, circa 1955

The first general purpose plastic charge cards were introduced in the 1950s by Diners Club, Carte Blanche and American Express. These weren't really "credit" cards because payment on the full statement was required every month and there was no revolving credit. In the early 1960s a plastic credit card named *BankAmericard* became popular and was accepted by most merchants. This was the beginning of the use of the modern credit card, with revolving credit. It later became the *Visa* card. In 1966 Master Charge (now *MasterCard*) was introduced.

In the late 1960s, there was a credit card scandal in which the credit card companies indiscriminately mailed tens of millions of unsolicited credit cards to people throughout the United States. The recipients included, among others, people who were bankrupt, unemployed, in prison, and also even infants and children.

Using a credit card in the early days was very different from today. Until the mid-1970s the credit card system was not computerized. Prior to that, when you used a credit card the cashier in most stores had to make a phone call to an authorization office. If approved, the credit card would be swiped over a layered paper transaction slip in a manually-operated mechanical credit card imprinter.

Bartizan Model CM2020 manual credit card imprinter

Many stores had "stolen card" books that were updated and distributed very frequently. These soft-cover books had lists of stolen credit card numbers, in numerical order. They also had lists of credit card numbers corresponding to cards used by "deadbeats." I remember that during the 1960s, when I would shop at a supermarket or a department store, the cashier would compare my credit card number with the listed numbers. I was glad to not be on any of the lists.

Telephone Order Shopping

Mail order shopping has been in existence for over a century. Purchasing goods that are advertised in a magazine or newspaper became popular in the mid 20th Century. It was common for the advertisement to have a portion for the purchaser to write in contact information, sever, and mail to the advertiser with a check or money order. Joe Sugarman, president of my client JS&A Group, had a better idea in 1970. It was to give customers an opportunity to purchase JS&A's goods over the telephone, using their credit card.

Prior to 1970, when goods were ordered over the telephone, they were either paid for on delivery or a check or money order was sent to the merchant. But Joe recognized the benefit of being able to order over the telephone with a credit card. Although it seems so obvious now, as billions of transactions are handled that way, it had not previously been done. Joe contacted the banks controlling the credit cards in the United States and convinced them to allow the use of credit cards for telephone purchases. He also obtained an 800 number from AT&T for the transactions.

It worked beautifully! Sales increased greatly with customers recognizing how easy it was to make a telephone-order purchase. Joe's idea opened up a whole new way to purchase goods, now referred to as "card-not-present" (CNP) transactions. Now CNP transactions usually involve purchases over the Internet, as well as telephone orders. Although there is a greater chance of fraud than with a card-present transaction, we have become accustomed to being asked for certain card details by the merchant. This may include cardholder's name, address, card number, expiration date, and security code.

AT&T was so thrilled with this use of an 800-number that they placed full-page advertisements in *The Wall Street Journal*, using JS&A's successful telephone order method as a sales tool for marketing 800-numbers.

Barcodes

Before the late 1970s, at the supermarket, the prices on everything were plainly marked on an attached price label, usually in the form of a sticker. There were no barcodes. If you went to a deli section, the clerk would cut the meat, wrap it and write down the price on the wrapper. If the price changed, a new price label would be used or the old price would be overwritten. Attaching price labels, typically with a label gun, was a major time-consuming chore in a supermarket.

After obtaining your groceries you would go to a check-out counter where the cashier would have a cash register. The cashier would view the price on each item and use the keys on the cash register to enter the price of each item, and then press the "Total" key. You would pay by cash and hopefully the cashier would give you the proper change from the cash drawer. It wasn't until the 1960's that a cash register could calculate the amount of change. Also, if the cashier was not adept at using the cash register keys the check-out could take a long time.

The use of barcodes on products, beginning in the late 1970's, provided a great advantage to the world of shopping. It was no longer necessary for the cashier to look for the price label on every item or for a store clerk to label and relabel every item. Simply scanning the barcode gave all the data needed regarding a product and the price and description could now be entered into a computerized cash register automatically by a simple scan of the barcode.

The earliest use of barcodes was not on consumer products. Barcodes were first used on railroad cars. I became aware of barcodes in the mid 1960s when I handled a trademark matter for the Santa

Fe Railroad relating to barcodes. The barcodes were in the form of colored reflective bars attached to the sides of the freight cars.

Freight car with attached *KarTrak* bar code
Photo credit: Trey Meador

The standard barcode used on railroad cars was developed by General Telephone and was named *KarTrak*. It was required on railroad cars throughout United States between the mid 60s and 1977. These barcodes on the sides of railroad cars were abandoned in 1977 because they were unreliable. The railroad car traveled at a high speed past a trackside scanner and the barcodes were often filthy, resulting in an inaccurate reading. Railroad cars are now tracked using a radio-based system.

Another early use of barcodes was by Federal Express and UPS in the early 1970s, for tracking packages. The United States Postal Service and DHS also track using barcodes. The barcode on the package is scanned multiple times throughout its journey. It's so simple to check the status of the delivery by using the tracking number. You can easily track a package on any smartphone or computer connected to the Internet. There's no doubt that delivery services and the public have benefitted tremendously by this great tracking system.

It wasn't until the mid-1970s that barcodes were used on consumer goods and groceries. The use of barcodes revolutionized the methods of pricing and inventorying. The container for the product contains a pre-printed barcode that is unique for each product. A twelve-digit UPC barcode is typically printed on groceries, while books carry a slightly different form, using the book's unique ISBN number.

A UPC barcode example

When checking out, the cashier can quickly scan the barcode on each item and a computerized terminal will produce a receipt that includes a description of each scanned item, its price, any applicable discount, the total balance, the cashier's ID, and other pertinent information. Data may be fed to an inventory storage medium to account for the removal of the items from inventory. By using barcodes there is no need to attach price labels on each of the products. The applicable prices can be displayed on an adjacent shelf and can be changed easily without relabeling each product.

Barcodes may take many forms and can be used for tracking almost anything. They can be used for tracking the location of people, rental vehicles, luggage, assembly line materials, hospital patients, documents, stored boxes, and the list goes on. Barcodes are commonly used on tickets. You can print your own theater tickets, transportation tickets, or other tickets having barcodes. You can store the tickets on your smartphone. At the appropriate venue the barcode will be scanned to determine the validity of the ticket.

There are numerous smartphone apps enabling barcode scanning. For example, the Amazon app has a barcode scanner. On many occasions I have seen an item at a store and then scanned the barcode with the Amazon app on my iPhone in order to compare prices. That is much faster than having to enter the product information on the Amazon app manually.

RFID Tags

One of the most important advances in tracking is the use of an RFID tag on any object you want to detect or follow. RFID stands for radio-frequency identification. RFID tags are miniature radio transmitters that use radio waves to transmit data to an RFID reader. Modern RFID tags comprise a base, a microchip integrated circuit, and an antenna. Some are so small they can hardly be seen.

An important use of RFID tags is in the alleviation of shoplifting. Many of the clothes and other items that we buy carry attached RFID tags. The tag is removed by the cashier when the item is purchased. As you leave the store you walk past the RFID reader posts which will sound a loud alarm if the RFID tag has not been removed. Lately many RFID store readers are hidden.

An RFID tag can take many forms, including stickers, key fobs, cards, etc. They are ideal for use in monitoring the movement of any object or person and for verifying such movement.

A popular use of RFID is in toll collection systems. For example, in E-ZPass, used in the northeast United States, a RFID transponder containing your account information is provided. The transponder is a battery-operated transmitter/receiver, placed on the front window of the vehicle. It transmits and receives data in the form of radio waves. An overhead antenna unit is located in a toll collection zone. When the vehicle passes under the antenna unit, the antenna unit sends radio waves communicating with the RFID transponder. Information from the transponder is read by the antenna unit and sent to a database, to deduct the toll from your account.

Do you have a contactless credit card? If so, the chip on it may include an embedded RFID tag. I've heard that it is possible for a hacker to approach you and use an RFID reader to obtain information from this type of credit card. Likewise, passports have embedded RFID tags, although they have to be open to be read. Wallets and passport holders are being sold that use shields to provide RFID blocking, avoiding this problem.

The Digital Wallet

Instead of having to carry credit or debit cards you can store the credit card and debit card information in your smartphone. This is commonly referred to as a "digital wallet." Two widely used digital wallets are Apple Pay and Google Pay.

I have Apple Pay on my iPhone and I use it occasionally. Since some merchants do not accept Apple Pay I also carry a credit card in my regular wallet. But Apple Pay has many extremely useful features. It's easy to install on your iPhone. Just go to Settings –Wallet & Apple Pay--Add Card, and enter the information by typing it or by using the camera on the iPhone. The information is not stored on the iPhone. It is encrypted and the encrypted data is stored in Apple's servers. Your credit card numbers are safe. Even a hacker can't get your information (unless you are hacked while you are sending card information to Apple from an unsecure Wi-Fi network).

When you use your digital wallet you don't need to hand or show a credit or debit card to a merchant. You just place your smartphone near the payment terminal and data is communicated via radio waves using a system known as NFC (near field communication). Instead of sending your credit card information (such as your credit card number) to the credit card network it sends a unique one-time random token, based on a secure data system called "tokenization."

Your digital wallet can "carry" a large number of cards. In addition to credit and debit cards, event tickets, coupons, boarding passes, and student ID cards can be stored. I usually use my iPhone wallet to store and display my boarding pass.

Although a digital wallet is considered by most to be a credit and debit card storage medium, it could also be linked directly to your bank account. It can be used for any financial transaction, creating an incentive for banks and all types of merchants to become involved.

One problem that I find frustrating is waiting on a checkout line while someone unfamiliar with Apple Pay or some other digital wallet attempts to complete a transaction. Often it ends up taking much longer than if they had retrieved and used a plastic credit card the old-fashioned way! However, I predict that the digital wallet is the wave of the future and that its use will flourish.

The ATM

Retail banking (i.e. consumer banking) goes hand-in-hand with buying things. One of the most significant innovations in retail banking occurred with the advent of the automatic teller machine (the "ATM").

A row of four ATMs at a Bank of America location, 2020

Prior to 1970, obtaining cash and/or depositing checks was often a hassle. I would go to a bank where I had an account, and prepare a "counter check" for the amount of cash that I wanted to withdraw and for the amount of the check deposit. I would get in a line to a teller's window. To finally arrive at the teller's window could take a substantial amount of time. I can remember waiting as long as a half hour to reach a teller, particularly on a payday such as a Friday.

In 1971 an ATM was installed at La Salle National Bank of Chicago, which was on the first floor of the building where my law firm was located. It was the first ATM in Illinois. It had a comic face attached and was called "MAC, THE FRIENDLY TELLER." I had recently opened a checking account there and I was offered an access card which I gladly accepted. I selected a PIN to use with the access card and began using it to withdraw cash from my checking account. I could not deposit checks at the time because the check deposit function on ATMs did not occur until 1980. However, the ATM was hardly ever being used by others and I enjoyed the convenience of not having to prepare a counter check and wait in a line for a teller to obtain cash. The ATM was located in the building lobby and was available for use anytime, so I didn't have to depend upon the bank being open.

As time went on, more customers began using the ATM and occasionally I had to wait for another customer to complete a withdrawal. Later, more ATMs were installed as they became more popular. La Salle Bank became a client of mine, and I registered their new service mark "MAC, YOUR FRIENDLY 24-HOUR BANKER."

By the mid-1970s ATMs were being installed in banks all over the world and in the following decade they could be found everywhere: in stores, airports, restaurants, along city streets, and wherever someone would want to obtain a cash withdrawal from his bank account. By the 1980s many of the ATMs accepted cash and check deposits, resulting in even less labor for bank tellers.

Independent ATMs that were not directly connected with banks were installed in smaller venues such as convenience stores. Most retail locations found that having an ATM increased foot traffic. Transactions made on a dedicated bank machine by a bank customer are usually free, but a non-customer is usually charged at least $3 by the bank per transaction. Transactions on an independent ATM have a surcharge that is charged by the owner of the ATM to the user.

The convenience of being able to withdraw cash and make deposits using a local machine is obvious. However, there is a dark side: the risk of being robbed. Withdrawing cash at night and/or in what appears to be a desolate area is risky. It's important to observe your surroundings and abstain from withdrawing cash if someone or something appears suspicious. Don't count the cash at the ATM. Make certain no one is following you from the ATM. If so, get to a busy area ASAP.

The use of skimmers on ATMs by thieves is another problem. A skimmer is a credit or debit card reading device that is attached by the thief to an ATM in order to capture the information from the card. If the card reader doesn't look or feel right it might be a skimmer. For example, it may look different from the card readers on other ATMs or it may feel loose. There is a greater chance that it will be found on an independent ATM which is generally serviced less than a bank's ATM. Lately thieves have used skimmers that are so small they can be hidden within the card reading slot of the ATM. An interesting short video about how ATM skimmers work can be viewed online at www.gerstman.com/skimmer.htm.

Don't think skimmers are only found on ATMs. They have been used on gas pumps, parking meters, and everywhere else a valid card reader is used.

Online Shopping

In my opinion, what we now consider to be "online shopping" began 1n 1995. Prior to 1995 there were networks that enabled online computer communication and the Internet was in existence. But there was no useful system enabling a consumer to view a product or service on his computer and purchase the product or service from the seller using his computer. This online shopping was made possible by the websites and web pages of the World Wide Web. It was the advent of the web browsers Netscape and Internet Explorer that opened up the world to online shopping.

At first, most businesses used their website to present their identity or their brand to the public. But it didn't take long for businesses to see that their products and services could be sold directly through their websites, using the purchaser's credit card. If telephone order worked, so could online ordering. Why not have an online retail store?

One of the earliest retail stores on the Internet was Amazon. com, founded by Jeff Bezos and introduced as a website to the public in 1995. At first, Amazon was solely a bookseller. I was an early customer. I was amazed at their large catalog of books, huge discounts and fast service. Amazon figured out how to make the purchase of books so convenient and inexpensive that it would become more efficient to purchase from Amazon than from a physical bookstore. Once you signed up as a customer there was practically no effort in making purchases. Your purchase would usually be delivered within a couple of days, which was much quicker than the typical mail or telephone order.

By 2000, Amazon had become a retailer of all types of products including music, video, electronics, housewares, hardware, toys, and its own e-book reader, the Kindle. What a wonderful way to send gifts without having to go shopping at a physical store!

Amazon made shopping even easier with its “1-click buying.” In 2005 Amazon introduced its Amazon Prime subscription service providing free and expedited shipping. I am an Amazon Prime member and one of the millions of Amazon customers whose address and credit card information are in an Amazon database. In this manner, once I find the item I want to buy I can simply click on a “Buy Now” button and the order is completed. That’s “one-click buying” in operation.

It’s usually very easy to find the item that you want to buy. You can simply enter the description in the search box of the Amazon website or the Amazon app on your smartphone. The Amazon app for the smartphone contains a bar code scanner that enables you to scan the barcode of the item that you are interested in and it will then be displayed on the Amazon app. This is particularly useful when you are in a store where you can scan the barcode of the item that you’re interested in for comparison shopping.

It’s hard to imagine the amount of inventory Amazon must have in its hundreds of millions of square feet of warehouse space throughout the world. In addition to having its own giant inventory Amazon provides fulfillment services for other sellers, and stores their inventories. About half of the inventory in Amazon warehouses is from other sellers, where it has been received, checked, scanned and stored. When an item from a third-party seller is purchased, Amazon retrieves the item from storage, processes the payment, packages the item, and ships and tracks the item. A fee is paid by the seller for these services.

I wondered how an item ordered from Amazon could be stored and found quickly in a giant warehouse having millions of items. The answer surprised me. The warehouse does not have designated areas for specific classes of goods. In other words, there is no camera section, soap section, shoe section, etc. Instead, an associate places each item in *any* nearby available bin that is suitable in size. It works like this: an associate removes the item from the box in which it was shipped to Amazon. The item's barcode is scanned, the scanner indicates the appropriate size of the bin to hold the item, the item is placed in any nearby available bin of the indicated size, and the bin barcode is scanned to correlate the item with the bin. When a customer orders the item, an associate will see the identification of the bin and will retrieve and process the item. In some of the warehouses the retrieving is done robotically.

I have never considered myself to be a retailer but I have sold some of my used art books on Amazon. Any individual or business can be a seller on Amazon. Amazon has become so popular that approximately half of all online retail sales are through Amazon.

Another type of market that is ideally suited for online sales is the auction market. Beginning in the 1960s I attended auctions and enjoyed bidding and sometimes "winning" various art and furnishings. Each auction had a very limited amount of auction lots. In 1995 the auction market changed monumentally when an online auction website named AuctionWeb (name later changed to "eBay") was introduced. eBay enabled every individual and business to list goods for auction and to bid on the goods that were listed. eBay received a great boost in 1997 when it became a primary source for Beanie Babies.

There are presently over a billion listings on eBay. Although at first every item was for sale only by auction, starting in 2000 sellers had the option of adding a "Buy It Now" button and a "Make an Offer" button. A buyer could avoid the auction mode and pay a fixed asking price to "win" the item immediately or the buyer could make an offer of an amount that is lower than the asking price.

Bidding on eBay can be a unique experience. eBay refers to the primary mode of billing as "automatic bidding." Here's how automatic billing works. You enter an amount that is the maximum amount that you want to bid on the item. Assume you want to purchase a camera that has a starting bid price of $250. For that amount, bid increments are five dollars. You are willing to pay a maximum of $400 so you enter the $400 amount and click "Place Bid." Your bid will be shown as $255 which is the $5 increment above the starting bid of $250. In short, eBay's automatic bidding raises the bid to the increment above the highest bidder's amount. If no one else bids in that auction the $255 will be the winning bid. However, if someone else bids $350 as their maximum bid, eBay will up the bidding amount to $355 and you are still the highest bidder. The other person will see that he is outbid and if he raises his maximum bid to $500, the bidding will be up to $405 and you are no longer the highest bidder. If no one had been against you from the beginning, even though you provided a maximum bid of $400, your winning bid is $255. Throughout the automatic bidding, all of the maximum bids are secret although you are advised when you are outbid.

Using eBay's automatic bidding system, bidders can constantly adjust their bids in response to their being outbid. To avoid being outbid this way, there are third-party programs and apps that enable what is called "sniping." There is a sniping app called *Bidslammer*. You login, enter the identification number of the item that you want to purchase on eBay, and enter the maximum amount that you are willing to bid. At about one second prior to the end of the auction of that item, *Bidslammer* automatically places a bid for you. A major benefit is that none of the other bidders see your bid until one second prior to the end of the auction. The bid is the next increment up from the previous bid, which may be lower than your maximum bid. During the auction, *Bidslammer* advises you if your maximum amount has been outbid by another bidder, giving you an opportunity to raise your maximum.

Now that you're the winning bidder on an eBay auction item that you've been seeking, how do you pay the seller? It's very simple. The seller states in his post what is acceptable. Most sellers favor PayPal, but many also accept credit or debit cards. You don't have to worry about the seller (who is usually an individual or small business) seeing your card numbers because the transactions are handled by eBay, which informs the seller by email when the payment is in the seller's account.

eBay formerly owned PayPal and the use of PayPal with eBay is extremely convenient. As soon as the buyer pays with PayPal the appropriate amount is immediately in the seller's account and both parties are notified. The seller is then required to ship the item. I like using PayPal for all eBay transactions because PayPal is considered very secure and has an excellent eBay buyer protection plan. More recently a buyer can also pay with Apple Pay or Google Pay, and eBay handles the transaction. In any event it is beneficial for the seller to act responsibly because eBay has a feedback system that allows the world to see the seller's feedback rating that has been provided by buyers.

Tickets to all types of events are purchased on eBay, sometimes at amounts that seem unbelievable. In 2007 eBay bought StubHub, a major online ticket broker. These online giants cover a large share of the resale ticket market and have made the purchasing of event tickets amazingly convenient. You can purchase tickets using a StubHub app on your smartphone, with a view of your seat location in the venue plus a view of the stage or field from your seat.

eBay has made the posting of items on its site incredibly easy. I use my iPhone for posting. I take the appropriate photos, fill in the necessary information, and then publish. The app guides you right through it. Posting tickets for sale on StubHub is so simple it can be accomplished in about a minute once you get used to it. The advent of online buying and selling is one of the greatest achievements I have seen in my lifetime.

EPILOGUE

I have been so fortunate to be able to observe the evolution of technology over the past 80 years. Moreover, I was lucky to have chosen a profession that enabled me, in fact required me, to view and understand diverse technological advances.

One of the most fascinating aspects of the tech evolution is the return of some of the traditional technology of the past that appeared to be obsolete. For example, many people are setting aside their digital cameras and are using vintage film cameras. Possibly it became cool because of attractive film cameras being used as props in so many ads and commercials. It's true that many film cameras were designed to be beautiful as well as functional. Film cameras have a long history and searching for and collecting them is fun. Also, I can attest as an old photographer that there is a special feeling to taking an analog photo with a film camera. The mechanical sounds, the clicks and whizzes that the camera makes are special. It poses a challenge. You have to select and frame your picture very carefully because you don't want to waste film and you won't see the pictures until they are processed. Many younger people are attracted to the use of the darkroom for chemically processing the film and watching the image emerge on the photo paper. And the entire picture-taking process for film does not involve the use of a computer or batteries.

Vinyl records have also returned, with a vengeance! One thing about playing digitized music from a compact disc, streaming, downloading, or from any other any digital storage medium – it's usually sharp, crisp and clean. When it's copied over and over it sounds the same when it's played. But maybe something is missing. You might not want the music sounding so crisp and clean. Maybe you want it sounding warm and mellow. Perhaps you want a vintage experience. That's where vinyl records come in.

There is the fun of going to a record store, many of which had closed but are now returning. You can buy a physical object having a groove containing the audio, as contrasted with streamed or downloaded digital data. Some people enjoy showing their record collection on social media. You may feel more involved in the music if you are handling the grooved record, placing it on an attractive solid turntable, carefully dropping the needle, and controlling a vacuum tube amplifier. The return of vinyl records has inspired the return of turntables and vacuum tube amplifiers. Some of the main music producers, including Sony Music, are back to selling vinyl records. For the first time since the 1980s, in 2020 more vinyl records were sold than CDs.

The significant advances of digital quartz watches have not impeded the market for traditional mechanical watches. In the 1970s, as quartz watches were becoming more popular with prices dropping sharply and watch functions increasing, some thought that the mechanical watch industry would end. Likewise, more recently some thought that smartwatches would destroy the mechanical watch industry. In both instances they were wrong. Sales of mechanical watches slipped at first but then recovered. Mechanical watch sales have recently flourished. Presently more than half of the amount spent by US consumers on watches is for mechanical watches.

What is the motivation for buying a mechanical watch? In contrast to a quartz watch a mechanical watch needs to be wound and is less accurate. Some quartz watches are designed to look like mechanical watches. But there is a certain status and glamour to having a mechanical watch. It makes a statement that you appreciate tradition. Fine mechanical watches are sold as an article of quality jewelry. They are ageless and can be passed on for generations, never requiring a battery. Different mechanical watch movements are still being developed making the mechanical watch industry both antique and contemporary.

There has been a revival in the sale of fountain pens. They require refilling with ink. They leak. But there is a certain feeling to having and using a fountain pen. They feel good to hold. The ink flows smoothly onto the paper without much pressure. Like with mechanical watches, they make a statement that you appreciate tradition. They endure and can be passed on to your children. They bring you back to simpler times.

Revivals of vintage devices may seem like the newest technology has been diminished. It hasn't. How can you fully appreciate the present and future technology if you disregard the past? I hope that readers of this book appreciate the devices and methods of the past, as well as the transformation to their present form. While these changes may appear to have occurred magically, they really didn't. Typically, engineers and scientists work to improve the past devices and methods, but change is usually incremental. In 1960, when I was employed as a patent examiner at the US Patent Office, I was amazed at first to find how most new inventions in the patent applications appear to be only slight improvements over earlier inventions. But these slight improvements add up and science and industry are advanced. Sometimes the most minor change in structure results in a synergy that provides a revolutionary new product. We, the consumers, are the beneficiaries.

INDEX

Symbols

A

B

C

D

E

F

G

Q

R

S

T

U

V

W

X

Y

Z

Lightning Source UK Ltd.
Milton Keynes UK
UKHW011431310521
384684UK00007B/821/J